The Open University

Mathematics: A Second Level Course

Topics in Pure Mathematics Units 4 and 5

GROUP AXIOMS

GROUP MORPHISMS

Prepared by the Course Team

The Open University Press

The Open University Press, Walton Hall, Bletchley, Buckinghamshire.

First published 1973.

Produced in Great Britain by
Technical Filmsetters Europe Limited, 76 Great Bridgewater Street, Manchester M1 5JY.

SBN 335 01362 7

This text forms part of the correspondence element of an Open University Second Level Course. The complete list of units in the course is given at the end of this text.

For general availability of supporting material referred to in this text, please write to the Director of Marketing, The Open University, Walton Hall, Bletchley, Buckinghamshire.

Further information on Open University courses may be obtained from The Admissions Office, The Open University, P.O. Box 48, Bletchley, Buckinghamshire.

1.1

Unit 4 Group Axioms

"The great notion of Group, . . though it had barely merged into consciousness a hundred years ago, has meanwhile become a concept of fundamental importance and prodigious fertility, not only affording the basis of an imposing doctrine—the Theory of Groups—but therewith serving also as a bond of union, a kind of connective tissue, or rather as an immense cerebro-spinal system, uniting together a large number of widely dissimilar doctrines as organs of a single body."

C. J. Keyser,
Lectures on Science, Philosophy and Art
(New York, 1908), p. 12

Contents

Group Cards

Group Cards and instructions for using them are bound in at the back of this booklet, after *Unit 5*.

Set Books

P. R. Halmos, *Naive Set Theory*, paperback edition 1972 (Van Nostrand Reinhold).
I. N. Herstein, *Topics in Algebra*, paperback edition (Xerox College/T.A.B.S., 1964).
B. Mendelson, *Introduction to Topology*, paperback edition 1972 (Allyn and Bacon).
M. L. Minsky, *Computation: Finite and Infinite Machines*, paperback edition 1972 (Prentice-Hall).

It is essential to have these books; the course is based on them and will not make sense without them.

Unit 4 is based on **Herstein**, Section 3 of Chapter 1 and Sections 1–5 of Chapter 2.

Conventions

Before working through this correspondence text make sure you have read *A Guide to the Course: Topics in Pure Mathematics*.

References to the Open University Mathematics Foundation Course Units (The Open University Press, 1971) take the form *Unit M100 3, Operations and Morphisms*.

4.0 INTRODUCTION

It is often the case that apparently diverse scientific phenomena do in fact share essential common properties. Much of modern mathematics is concerned with the recognition of such properties and the formulation of mathematical systems which model them. Unencumbered by the irrelevant detail in the original context, the basic structure of the mathematical system often becomes apparent and the system lends itself to a clear and extensive analysis. Such is the case with the algebraic system discussed in this unit.

A group is a set of elements taken together with a binary operation which is required to satisfy four basic axioms. The virtue in studying such a simply-defined system lies in the fact that the theory is applicable to a wide variety of situations. The success achieved in the past by the application of groups in important and exciting physical and mathematical contexts indicates the wide-ranging relevance of group theory.

In this unit, we discuss first a mathematical system which is already known to us—the integers. Because many of the constructions involved in our study of groups are exemplified by the integers and their operations, this system will be a valuable aid in our preliminary investigation of group theory. We then present the definition of a group and the immediate elementary consequences of that definition. Lastly, we introduce the concept of subgroup and investigate some results involving subgroups. Much of the group theory in this unit was considered in *Unit M100 30, Groups I* and *Unit M100 33, Groups II*, and we assume that you have some familiarity with these units. However, the approach taken here is significantly different from that in the Foundation Course. Rather than dwelling on specific groups of given orders, we concentrate more on the abstract group concepts, and so obtain, not only the general structural properties common to all groups, but also the elementary means for constructing groups.

Although **Herstein** is written in a rather concise manner, you should not experience any great difficulty in reading the text in conjunction with the notes in this unit. However, you may find it difficult to visualize an approach to the problems posed at the ends of the sections. Although the solutions to these problems are usually short and of a mechanical nature, they are highly non-algorithmic. That is, you will not very often discover a proof technique which can be applied automatically in several problems. Nevertheless, the material needed for the problems always lies in the text preceding the problems. Therefore, if you are having trouble solving problems, we suggest that you exercise a little patience. We shall do our best to motivate the proofs, and, as the course develops, you will find the questions easier to answer. It will be a slow process, however, because the mathematical education you have obtained prior to this course has been mostly concerned with arithmetic, symbolic algebra and classical geometry, whereas the sort of algebra we are learning now is concerned with the manipulation of *ideas* rather than symbols. By the end of this unit, you will have obtained a fuller appreciation of the distinction between the traditional algebra of school years and the abstract algebra of group theory.

Aims

After working through this text, you should be able to:

(i) understand (for the integers) the concepts of greatest common divisor, relatively prime, and congruence modulo n;

(ii) be able to manipulate and use in proofs the equations expressing the assertion that two integers a and b are relatively prime;

(iii) be able to construct the equivalence classes of the integers modulo n, for any integer n, and have some facility with the binary operations of addition and multiplication defined on these equivalence classes;

(iv) state the axioms of a group and be able to show explicitly whether or not a given mathematical system is a group;

(v) determine whether or not a given subset of a group is a subgroup;

(vi) understand the decomposition of a group into cosets and know how to find those cosets in a given situation;

(vii) understand Lagrange's Theorem and its important applications in proofs using counting techniques;

(viii) enjoy a deeper awareness of the basic structure of groups and the use of abstract arguments in solving problems.

4.1 THE INTEGERS

4.1.0 Introduction

In the first three units of this course, we travelled a lengthy (and possibly tortuous) path from the early set axioms through many ideas fundamental to mathematics to the set ω of natural numbers, the arithmetic operations defined on that set, and an ordering of the set. We know from *Unit M100 34, Number Systems* that from ω we can construct the set Z of integers, together with its arithmetic operations of addition, subtraction, and multiplication, and also an order on Z. As we indicated in that unit, we can then continue by constructing the rational numbers and the real numbers. Comforted by the knowledge that the "normal" arithmetic that we do every day has a philosophical basis in naive set theory, we now return to normality. As stated in the first reading passage from **Herstein**, we now assume a basic knowledge of the integer operations, and we recall only infrequently the set-theoretic notions involved with them.

We shall adopt Herstein's notation and use J instead of Z for the set of integers. (We shall need Z later to denote a special group.)

4.1.1 Divisibility

We shall not ask you to work through the first sixteen pages of **Herstein**. Much of the material there has been covered in the Foundation Course and in the first three units of this course. As necessary in subsequent work, we shall draw your attention to the minor differences in notation and to any results that may be expressed in an unfamiliar manner.

*READ **Herstein**: page 17, Section 3 to page 18, line -13 (end of Corollary).*

Notes

(i) **Herstein**: *page 17, line 19.*
   The definition of $|a|$ is

$$|a| = a \qquad \text{if} \qquad a \geqslant 0,$$
$$|a| = -a \qquad \text{if} \qquad a < 0.$$

(ii) **Herstein**: *page 17, line -18.*
   Notice how, at the outset, we assume some of the "elementary" facts about the integers, facts which by Halmos' standard are very profound indeed. This is typical of creative mathematics: when one mathematician communicates with another in a book or a paper, he assumes those results which he thinks he and his reader know or ought to know. There is no attempt to create the sort of logical order from basic principles as in **Halmos**. There is no need to resort to such a lengthy exercise, because we are confident that it can be done.

(iii) **Herstein**: *page 17, lines -14 to -12.*
   Since Herstein does not present the axioms, we must constantly make assumptions about what we are expected to know and what requires justification. We hope, by example and the occasional remark, to help create a feeling for what is a satisfactory verification of a particular result. As a demonstration, consider Herstein's remark that, if

$$b|g \qquad \text{and} \qquad b|h,$$

then

$$b|(mg + nh),$$

for arbitrary integers m and n. An acceptable verification is:

$$b|g \Rightarrow g = kb,$$
$$b|h \Rightarrow h = lb,$$

for some integers k and l. Therefore,

$$mg + nh = mkb + nlb$$
$$= (mk + nl)b,$$

so that $b|(mg + nh)$. Among the properties of the integers used in this proof are associativity of multiplication, closure of multiplication and addition, and distributivity of multiplication over addition, but we have not explicitly stated this in the proof.

(iv) **Herstein**: *page 17, line −2 to page 18, line 2.*
This technique is often used when proving the uniqueness of a mathematical object satisfying given conditions. That is, we show that any two objects satisfying the given conditions must be equal. The conclusion is that all objects satisfying the given conditions are indistinguishable, and thus the object is unique.

(v) **Herstein**: *page 18, line 5* (*statement of Lemma 1.5*).
To illustrate this lemma for yourself, choose a pair of integers a and b, write out some members of the set $\mathfrak{M}$, and verify that the smallest positive member of $\mathfrak{M}$ is (a, b).

(vi) **Herstein**: *page 18, line 11.*
Note that Herstein uses here the fact that every non-empty set of natural numbers (in this case, a subset of $\mathfrak{M}$) has a least element. This property was discussed in *Unit 3, Sets and Numbers*, Section 3.3.2.

General Comment

Lemma 1.5 and its corollary are extremely important and will be used in several proofs in our group theory. The theorem

$$\text{if } (a, b) = d, \text{ then } \exists m_0, n_0 \in J \text{ such that } d = m_0 a + n_0 b$$

is an example of an *existence theorem*. We may derive useful results simply by *knowing that m_0 and n_0 exist*.

Note that the converse of this theorem is false. For example, if we know that

$$ma + nb = 6,$$

it does *not* follow that $(a, b) = 6$. It *does* follow that any number which divides a and b also divides 6. In particular,

$$ma + nb = 6 \Rightarrow (a, b)|6.$$

Notice, however, that the converse of the corollary is true:

$$ma + nb = 1 \Rightarrow (a, b)|1$$

and therefore

$$ma + nb = 1 \Rightarrow (a, b) = 1.$$

SAQ 1

Write an outline of the steps in the proof of Lemma 1.5 in the order given by Herstein.

(Solution is given on p. 29.)

SAQ 2

The positive integer d is said to be the *least common multiple* of the non-zero integers a and b if

(1) $a|d$ and $b|d$;

(2) whenever $a|x$ and $b|x$, then $d|x$.

Show that d exists and is unique.

(HINT: Begin by considering the set

$$S = \{s \in J : s > 0, a|s \text{ and } b|s\},$$

and bear in mind the name we have given to d.)

(Solution is given on p. 29.)

4.1.2 Prime Numbers

*READ **Herstein**: page 18, line −12 to the bottom of page 20.*

Notes

(i)   ***Herstein***: *page 18, lines* −*11,* −*10.*
If the number 1 were not excluded from the primes, the factorization of natural numbers into a product of primes would not be unique, for we could then include an arbitrary number of 1's among the factors.

(ii)   ***Herstein***: *page 19, line* −*18 to page 20, line 3.*
The version of induction that Herstein is using here is simply "course of values induction" that we mentioned in *Unit 3*, Section 3.2.1, Note (i). It is true that we can deduce both forms of induction from the property that every non-empty set of positive integers has a minimal element. However, in *Unit 3* we took the Principle of Mathematical Induction as an *axiom*, i.e., one of the fundamental properties of ω, and derived from it the minimal element property. See *Unit 3*, Section 3.3.2, Note (vii), where we established the Well Ordering Principle.

(iii)   ***Herstein***: *page 20, lines 4 to* −*1.*
Again, for the proof of uniqueness, we require the stronger form of the induction premises, namely "if $P(u)$ for all $u : 2 \leqslant u < a$, then $P(a)$" (***Herstein***: *lines 12, 13*).

Notice that Herstein (*line 6*) arranges both sets of factors in descending order of magnitude; this is just to make the factorization *unique*.

To prove that $\alpha_1 = \beta_1$ (*lines* −*15 and* −*14*), assume that $\alpha_1 \geqslant \beta_1$. (This again is only a matter of naming.) Then

$$a = p_1^{\beta_1}(p_1^{\alpha_1 - \beta_1}P) = p_1^{\beta_1}Q,$$

where

$$P = p_2^{\alpha_2} \cdots p_r^{\alpha_r} \quad \text{and} \quad Q = q_2^{\beta_2} \cdots q_s^{\beta_s},$$

and therefore

$$p_1^{\alpha_1 - \beta_1}P = Q.$$

But $p_1 > q_2 > \cdots > q_s$, since $p_1 = q_1$. Consequently Q contains no factor p_1 and therefore $p_1^{\alpha_1 - \beta_1} = 1$, that is, $\alpha_1 = \beta_1$.

SAQ 3

Prove the Corollary in ***Herstein***: *page 19, line 9.*
(HINT: Prove the result first for the product $a_1 a_2$. Then prove the result for any product $a_1 a_2 \cdots a_n$, by induction.)

(Solution is given on p. 29.)

SAQ 4

Herstein: *page 23, Problem 9.*

(Solution is given on p. 30.)

4.1.3 Congruences

READ **Herstein**: page 21, line 1 to page 22, the end of Section 3.

Notes

(i) **Herstein**: page 21, line -10 to line -9.

Although there are only n distinct congruence classes of the integers modulo a given integer n, there are an infinite number of ways to represent each class. For example, for the integers mod 5, we have

$$\cdots = [-4] = [1] = [6] = [11] = \cdots$$

The congruence class $[a]$ mod n is usually represented by $[k]$, where $[k] = [a]$ and $0 \leqslant k < n$.

(ii) **Herstein**: page 22, lines 6 to 10.

Note that, since $[i]$ and $[j]$ are *not* integers, but equivalence classes, we must actually define "addition" and "multiplication" for these classes. We may think of this in terms of morphisms. The function

$$f : i \longmapsto [i] \qquad (i \in J)$$

maps the set of integers J onto the set of congruence classes J_n. That is, f is the *natural mapping*, defined in *Unit M100 19, Relations*, Section 19.2.2.

We have a closed binary operation $\circ$ in J ($\circ$ could be either addition or multiplication) and we wish to define a closed binary operation $\square$ in J_n so that f is a morphism. Thus, we demand that

$$f(i \circ j) = f(i) \square f(j).$$

That is, we want the commutative diagram:

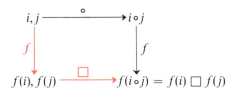

From statement (3) of Lemma 1.7, we know that addition and multiplication are compatible with the function f, since whenever

$$[i] = [i'] \quad \text{and} \quad [j] = [j'],$$

we have

$$[i \circ j] = [i' \circ j'].$$

We then know that we can *induce* a binary operation $\square$ on J_n by defining

$$f(i) \square f(j) = f(i \circ j), \qquad \text{i.e., } [i] \square [j] = [i \circ j].$$

(iii) **Herstein**: page 22, line 13 to line 19.

Equations (1) to (7) follow from the corresponding properties of the integers and the compatibility results for addition and multiplication.

SAQ 5

Herstein: page 23, Problem 13.
(Assume that $n > 1$. Use the corollary on **Herstein**: page 18.)

(Solution is given on p. 30.)

SAQ 6

Herstein: page 23, Problem 17.
(Assume that $n > 1$.)

(Solution is given on p. 30.)

4.2   DEFINITION OF A GROUP

4.2.0   Introduction

READ Herstein: *page 25, line 1 to page 26, line 11.*

Notes

(i)   *Herstein*: *page 25, line 7.*
By the expression "one-operational systems", Herstein means mathematical systems endowed with one binary operation. The set of integers together with the binary operation of addition is an example of a one-operational system.

(ii)   *Herstein*: *page 25, lines 9 and 10.*
Homomorphisms were introduced in *Unit M100 3, Operations and Morphisms*, and used throughout the Foundation Course. Quotient construction was introduced in *Unit M100 19, Relations* and used in *Unit M100 34, Number Systems* and *Unit M100 36, Mathematical Structures*.

4.2.1   Notation and Terminology

Each of the concepts developed in the next section of **Herstein** was discussed in the Foundation Course. There are, however, some differences in terminology, so we present a brief description of Herstein's definitions.

For Herstein, the terms *mapping* and *function* are synonymous; the corresponding term in the Foundation Course was *function*. In the great majority of instances, Herstein uses the term *mapping*. If σ is a mapping from S into T and $t \in T$ is the image of $s \in S$ under σ, then Herstein (*page 10, last paragraph*) denotes this fact by writing the mapping on the right :*

$$t = s\sigma.$$

This practice is used by many algebraists. Also, if σ is a mapping from S into T and τ is a mapping from T into U, then the composite of σ and τ is also called their *product* and it is the mapping from S into U expressed, for every $s \in S$, by $(s\sigma)\tau$ or $s(\sigma\tau)$ or simply $s\sigma\tau$, that is, reading from *left to right*, perform first σ, then τ. (This would have been expressed as $\tau \circ \sigma(s) = \tau(\sigma(s))$ in the Foundation Course.) The present terminology can be represented by the following commutative diagrams.

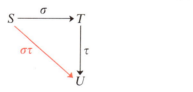

For a given non-empty set S, finite or infinite, Herstein denotes by $A(S)$ the set of all one-to-one mappings from S *onto* S. These mappings are called *permutations* on S.

In *Unit M100 30, Groups I*, Section 30.2.6, we showed that $(A(S), \circ)$ is a group. This is the example with which Herstein begins the next reading passage.

---

* In later units we shall find that Herstein is not always consistent in his notation.

4.2.2 The Axioms

In deciding which axioms are to be included in the definition of a group, we consider not only the great number of mathematical structures which satisfy the proposed axioms, but also the amount of useful theory that develops logically from those axioms. In other words, the final choice of the axioms is a result of a trial and error process, performed by mathematicians over many years. In the case of groups, the mathematical richness and the extensive applicability of the theory testify to the appropriateness of the axioms. Later, we shall study other important structures whose axiom systems include only some of the group axioms, and yet additional structures whose axioms properly contain the group axioms.

READ Herstein: Section 1, page 26 to page 27.

Notes

(i) *Herstein: page 26, lines 18 to 20* (property (1) of $A(S)$).
The product ∘ is a mapping of $A(S) \times A(S)$ into $A(S)$ and thus ∘ is a *closed binary operation*.

(ii) *Herstein: page 26, lines −12 to −10*.
Read *Herstein: page 15, lines −14 to −6*, where two such elements are constructed.

(iii) *Herstein: pages 26 and 27* (definition of a group).
This definition is an example of the process of abstraction. Although the structure of $(A(S), ∘)$ is a fairly general one, we go a step further and extract from this structure four important properties. We then use these properties to define a new and completely abstract concept, of which $(A(S), ∘)$ is an example. It should be emphasized that the four group axioms are properties of the set G and the binary operation ∘ *together*. It is the pair $(G, ∘)$ which satisfies the axioms and thus forms the group. Herstein uses the generally accepted notation of referring to G as a group whenever the binary operation with which we are concerned is understood. We shall henceforth adopt the latter notation and denote a group by the letter G, unless there is some ambiguity concerning the operation, in which case we shall revert to the Foundation Course notation, and use $(G, ∘)$.

(iv) *Herstein: page 27, line 8*.
Cayley's Theorem was discussed briefly in *Unit M100 30, Groups I*, Section 30.2.6.

(v) *Herstein: page 27, lines 9 to 14*.
On first reading, this passage may appear to indicate that all groups with more than two elements are non-Abelian. That assertion is untrue. Even though $A(S)$ is non-Abelian whenever S contains three or more elements, $A(S)$ contains subsets which, together with the operation ∘, are themselves groups; some of these latter groups are Abelian.

(vi) *Herstein: page 27, line −15*. (Proof that $o(A(S)) = n!$).
This is perhaps an obvious assertion. Nevertheless, consider the slightly more general assertion: if S and T each contains n elements, then the total number of one-to-one mappings from S onto T is $n!$ We shall prove this by induction on n, and the desired result, $o(A(S)) = n!$, will follow by taking $T = S$.

If S contains just one element, s_1, and T contains just t_1, then the only one-to-one mapping from S onto T is σ such that $s_1\sigma = t_1$. Since $1! = 1$, we have completed the first step of the induction. Now, suppose the assertion is true for all pairs of sets, each containing the same number of elements, this number being less than or equal to n. Let $S = \{s_1, s_2, \ldots, s_{n+1}\}$ and $T = \{t_1, t_2, \ldots, t_{n+1}\}$. Consider first the set of one-to-one mappings from S onto T in which the image of s_{n+1} is t_j for some fixed j. The restriction of each such mapping to $S - \{s_{n+1}\}$ is a one-to-one mapping from $S - \{s_{n+1}\}$ onto $T - \{t_j\}$. Since each of the latter sets contain only n elements, it follows from the induction hypothesis that there are altogether $n!$ distinct one-to-one mappings from $S - \{s_{n+1}\}$ to $T - \{t_j\}$. That is, there are altogether $n!$ distinct one-to-one mappings from S onto T which map s_{n+1} onto t_j.

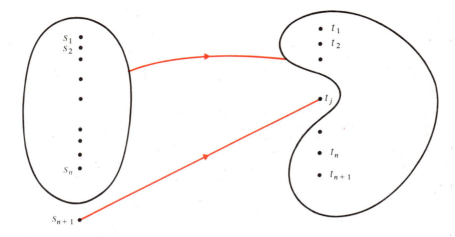

Considering all the possible images of s_{n+1} in T, that is, taking $j = 1, 2, \ldots, n + 1$ in turn, we obtain all the distinct one-to-one mappings from S onto T, a total of $(n + 1)n! = (n + 1)!$ mappings. This completes the induction step and the proof.

SAQ 7

Which of the following $(G, \circ)$ are groups?

(i)   (rationals $- \{0\}$, multiplication)

(ii)   (reals, multiplication)

(iii) $(J_n$, addition mod n)

(iv) $(J_n - \{[0]\}$, multiplication mod n)

(v) $(\{[a] \in J_n - \{[0]\} : (a, n) = 1\}$, multiplication mod n)

(Solution is given on p. 31.)

4.3 EXAMPLES AND ELEMENTARY PROPERTIES

4.3.0 Introduction

At this stage, the study of a few examples of groups, both finite and infinite, is an instructive exercise. Group structures abound in mathematical systems which are already familiar to us, and our knowledge of these systems will enable us to predict some of the basic properties common to all groups. This prediction process is of somewhat limited use, however, for it is sometimes difficult to recognize the salient features of a structure amidst all the accompanying detail of that structure. After all, any theorem on groups is a logical consequence of only four group axioms. So, despite the fact that some interaction between example and theory can be useful, especially at the beginning of a theory, it is also important to become accustomed to dealing with groups as abstract concepts.

4.3.1 Examples

READ Herstein: Section 2, pages 27 to 29.

Notes

(i) *Herstein: page 27, Example 2.*
For small finite groups such as this one, it is convenient to display the table of products:

$\times$	1	-1
1	1	-1
-1	-1	1

(See *Unit M100 30, Groups I.*)
The table shows clearly that G is closed under multiplication, that 1 is an identity element, that each of 1 and -1 is its own inverse and that G is Abelian. Associativity can be verified by considering each of the eight cases $1 \times (1 \times 1)$, $1 \times (1 \times -1), \ldots, -1 \times (-1 \times -1)$ separately, or by noting that G is a subset of the integers, so that associativity of G is inherited from the entire set of integers under multiplication.

(ii) *Herstein: page 27, line -2.*
The product in S_3 is, of course, composition of mappings.

(iii) *Herstein: page 28, lines 1 to 3.*
Notice that this *definition by induction* is a straightforward application of the Recursion Theorem (*Halmos: page 48*). The application to a^k can be expressed as follows. For any given $a \in G$, let f be the function from G into G defined for each $g \in G$ by

$$f(g) = a \cdot g.$$

The theorem states that there exists a function u from ω into G such that $u(0) = e$, the group identity element, and

$$u(k + 1) = f(u(k)) = a \cdot u(k),$$

for all $k \in \omega$. For each k, $u(k) \in G$ because it is the product of two elements from G. Since $u(k)$ is just the product of a by itself k times, Herstein denotes $u(k)$ simply by a^k.

To obtain the elements of G that Herstein denotes by a^{-k}, we need only replace a in the above definition of $u(k + 1)$ by a^{-1}, which is an element of G, by group axiom (4).

It is not usual to go back to the Recursion Theorem for such simple definitions. We did so in this case to demonstrate that it can (and, at least once, should) be done.

(iv) **Herstein**: *page 28, lines 6 and 7.*

Equations (1) and (2) will enable us to economize both our thought and our writing in further work. The proofs of the results involve an elementary application of induction, and we give the proof only for Equation (1) for $m \geqslant 0$ and all n. Fix n to be any integer. For $m = 0$,

$$a^0 \cdot a^n = e \cdot a^n \qquad \text{(definition of } a^0)$$

$$= a^n \qquad \text{(identity)}$$

$$= a^{n+0} \qquad \text{(integer arithmetic).}$$

Now, suppose (1) holds for $m = 0, 1, \ldots, k$. Then

$$a^{k+1} \cdot a^n = (a \cdot a^k) \cdot a^n \qquad \text{(definition of } a^{k+1})$$

$$= a \cdot (a^k \cdot a^n) \qquad \text{(associativity)}$$

$$= a \cdot a^{k+n} \qquad \text{(induction hypothesis)}$$

$$= a^{k+n+1} \qquad \text{(definition of } a^{k+n+1})$$

$$= a^{(k+1)+n} \qquad \text{(integer arithmetic).}$$

(v) **Herstein**: *pages 27–29, Example 3.*

Notice again that writing $\phi \cdot \psi$ for the composite operation "ϕ followed by ψ" is a consequence of Herstein's convention (which is the usual convention among group-theorists) that if a set S is mapped to itself by an operation ϕ, and ϕ maps an element s of S to the element t, then we write $t = s\phi$ and $t\psi = s(\phi\psi)$ (or $s\phi\psi$). In the present example, however, we are not concerned with the elements of the set S but only with the mappings (the mappings are the elements of the group).

The effect of $\phi \cdot \psi$ can be visualized by following the arrows in the diagram:

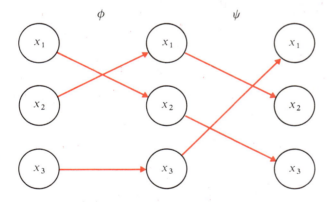

At this stage you may find it helpful to examine the GROUP CARDS. The cards, together with instructions for labelling and using them, are bound in at the back of this booklet, after *Unit 5*. We suggest that you label the cards before proceeding.

Using the group cards, you will reduce the above diagram to

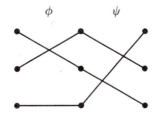

which shows that the composite operation $\phi \cdot \psi$ is represented by the card

$$\phi \cdot \psi$$

In the text we shall not often make explicit reference to the cards, but you may find the pattern of paths across a sequence of group cards enables you to get a clearer picture of the composition of a sequence of elements of a group—and may well find the manipulation of the cards a short cut to group-arithmetic.

For example, one way of representing the group manipulations in Herstein's Example 3 is the following:

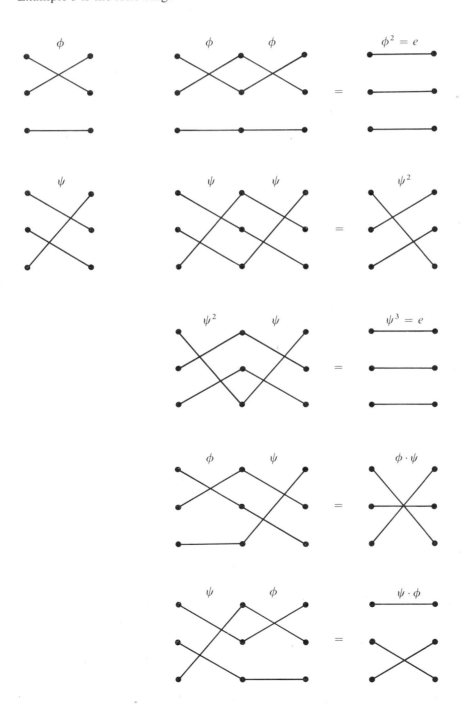

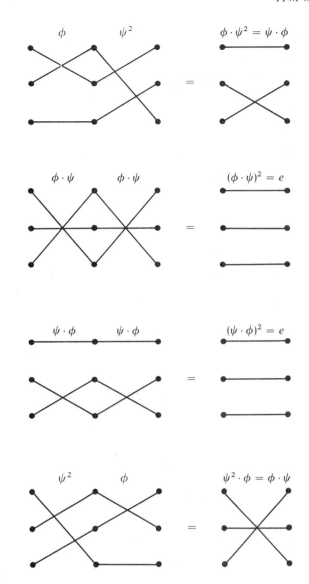

(vi) **Herstein**: *page 29, Example 4.*

Cyclic groups are relatively important in the study of group structures. Closure follows from the definition of a^{i+j}; associativity is easily verified by using the first exponent rule; $a^0 = e$ is obviously the identity; the inverse of a^i is a^{n-i}. One way to see why the group is called cyclic is to consider the consecutive increasing powers a^i, $i = 0, 1, 2 \ldots$ and to note that they produce in order the group elements $a^0, \ldots, a^{n-1}$ periodically.

A "geometric realization" of a group G is a set of transformations of a geometrical object to itself, which, together with the composition of such transformations, forms a group H isomorphic to G. In actual fact, we are discussing several levels of abstraction in this example:

(1) the general concept of a cyclic group of order n;
(2) an algebraic description of that group;
(3) an example (from geometry) of a cyclic group of order n.

To verify that H represents G, we should show that the mapping given by

$$e \longmapsto e$$

$$a^i \longmapsto \rho_n^i \qquad i = 1, 2, \ldots, n - 1,$$

is an isomorphism. Extensive discussions of isomorphisms between groups occur in later units of this course, so we shall not consider them in detail here.

Note that the finite set

$$H = \{e, \rho_n, \rho_n^2, \ldots, \rho_n^{n-1}\}$$

is a subset of the infinite set of rotational symmetry operations, K say, which in turn is a proper subset of the much larger set $A(S)$; each of H, K and $A(S)$, together with the operation of composition, is a group:

$$(H, \circ) \subset (K, \circ) \subset (A(S), \circ).$$

SAQ 8

Two elements of a group are represented by:

Verify that

(i) $(\phi \cdot \theta \cdot \phi)^2 = e$

(ii) $(\theta \cdot \phi)^2 = \phi \cdot \theta$

(iii) the set $\{e, \theta, \phi, \theta \cdot \phi, \phi \cdot \theta, \theta \cdot \phi \cdot \theta\}$, with the operation of composition, forms the group S_3.

(Solution is given on p. 32.)

SAQ 9

Give an example of a cyclic group of order n which has already been discussed in this unit.

(Solution is given on p. 33.)

4.3.2   Preliminary Lemmas

In this section, Herstein proves some basic and useful theorems that follow immediately from the group axioms. The proofs, which use only elementary logic, are concise yet easy to follow. However, the manipulation involved in subsequent group
theory will be made considerably easier once we have recognized the existence and
applications of the facts presented in these theorems. Again, most of the results were
proved previously in *Unit M100 30, Groups 1.*

READ Herstein: Section 3, pages 29 and 30.

Notes

(i)   *Herstein: page 30, lines 11 to 13.*
These basic computational tools are often referred to as the *cancellation laws*
(left and right, respectively). You will find yourself using them with group
products as often as you do with integer products.

(ii)   *Herstein: page 30, lines 15 and 16.*
Using the group cards for S_3, we can show that $\phi \cdot \psi = \psi^{-1} \cdot \phi$, so that certainly
$a \cdot x = y \cdot a$, but $\psi \neq \psi^{-1}$, that is, $x \neq y$.

Note that, in an Abelian group,

$$a \cdot x = y \cdot a \Rightarrow a \cdot x = a \cdot y \Rightarrow x = y.$$

(iii)   *Herstein: page 30, Lemma 2.2.*
The solution of $a \cdot x = b$ is $x = a^{-1} \cdot b$, since

$$a \cdot x = b \Leftrightarrow a^{-1} \cdot (a \cdot x) = a^{-1} \cdot b$$
$$\Leftrightarrow (a^{-1} \cdot a) \cdot x = a^{-1} \cdot b$$
$$\Leftrightarrow \qquad e \cdot x = a^{-1} \cdot b$$
$$\Leftrightarrow \qquad x = a^{-1} \cdot b.$$

SAQ 10

Herstein: page 31, Problem 2.
(HINT: Use proof by induction.)

(Solution is given on p. 33.)

SAQ 11

Let G be a group. Prove that G is Abelian if and only if, for all $a, b \in G$,

$$(a \cdot b)^2 = a^2 \cdot b^2.$$

(Solution is given on p. 33.)

SAQ 12

Herstein: page 31, Problem 8.
(HINT: Let $o(G) = n$; consider first any element $a \in G$, and investigate the $n + 1$
elements $a^0, a^1, \ldots, a^n$.)

(Solution is given on p. 33.)

4.4 SUBGROUPS

4.4.0 Introduction

Given a group G, there exist subsets of G which, together with the product in G, are themselves groups. In S_3, for instance, a brief search yields the subsets

$$T = \{e\},$$
$$H = \{e, \phi\},$$
$$N = \{e, \psi, \psi^2\},$$

each of which forms a group. The existence of these "subgroups" has immediate applications. For a given group, we can learn more about its structure by examining the number and form of its subgroups. Conversely, from any group G, it is possible to construct other groups containing G as a subgroup. By thus considering groups to be building materials for other groups, we shall be able to describe techniques for categorizing and analysing groups.

4.4.1 Definition and Recognition

READ Herstein: page 32, beginning of Section 4 to page 34, line 5.

Notes

(i) *Herstein: page 32, lines -18 and -17.*
In other words, the relation "is a subgroup of" on a set of groups, is transitive. This follows from the transitivity of set inclusion and the fact that the product is the same in all three groups. In S_3, for example, this is demonstrated by the subgroups T, H and S_3 respectively, and by T, N and S_3.

(ii) *Herstein: page 32, Lemma 2.3.*
This lemma states a *minimal* set of requirements on H for it to be a subgroup of a *non-finite* group G. For instance, the lemma would still be correct if we added the criterion that the product in H should be associative. There is no necessity to do so, however, because associativity in G automatically implies associativity in H. Note, however, that if H is to be a subgroup of G, we *must* have (1) *and* (2) for although ab and a^{-1} exist in G, they could otherwise be in $G - H$.

However, note that we can dispense with (2) if G is a *finite* group (Lemma 2.4).

(iii) *Herstein: page 33, proof of Lemma 2.4.*
The arguments used here occur *again and again* in the discussion of finite groups.

(iv) *Herstein: page 33, line 8.*
That is,

$$a^r = a^s \Rightarrow a^{r-s}a^s = a^s$$

and $a^{r-s} = e$ is obtained by cancelling a^s.

(v) *Herstein: page 33, Example 1.*
Verify that H is a subgroup. Notice that H is infinite, so you have to show that it contains its inverses, as well as showing closure.

We shall prove in SAQ 13 that $H_n \cap H_m$ is also a subgroup of G, and its form will be discovered in SAQ 14. In the meantime, note that $g \in H_6 \cap H_9$ if and only if $g = 6k$ and $g = 9l$ for some integers k and l. That is,

$$H_6 \cap H_9 = \{s \in J : 6|s \text{ and } 9|s\}.$$

(vi) *Herstein: page 33, Example 2.*
Remember that, since ϕ is a mapping, the expression $x_0\phi$ denotes the image of x_0 under $\phi \in A(S)$. $H(x_0)$ is the set of mappings in $A(S)$ under which x_0 is invariant.

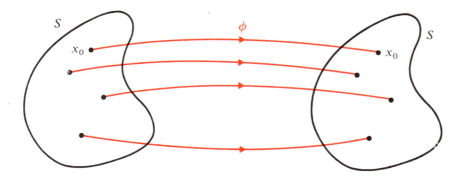

To show that $H(x_0)$ is a group, we use Lemma 2.3. If $\phi, \psi \in H(x_0)$, we have that

$$(x_0\phi)\psi = x_0\psi = x_0$$

so that $\phi\psi \in H(x_0)$. Also, if e is the identity mapping in $A(S)$,

$$x_0 = x_0 e = (x_0\phi)\phi^{-1} = x_0\phi^{-1},$$

and thus $\phi^{-1} \in H(x_0)$.

$H(x_0) \cap H(x_1)$ is just the set of mappings which leave both x_0 and x_1 invariant.

(vii) **Herstein**: *page 33, Example 3.*

Here we shall depart from Herstein's notation. Parentheses are used in so many different contexts that we prefer to denote the cyclic subgroup generated by a by

$$\langle a \rangle,$$

rather than (a). For future reference, we say that an element a of a cyclic group G is a *generator* of G if $\langle a \rangle = G$. For example, if G is the cyclic group of order 6, with elements

$$e, a, a^2 = p, a^3 = q, a^4 = r, a^5 = s,$$

then $\langle a \rangle = \langle s \rangle = G$, but $\langle p \rangle, \langle q \rangle, \langle r \rangle$ are cyclic groups of orders 3, 2, 3 respectively.

The fact that Abelian groups are not necessarily cyclic is demonstrated by the *Klein 4-group*. (See *Unit M100 30, Groups I*, Section 30.2.4.) It is Abelian but not cyclic, because $x^2 = e$ for each of its elements x. The elements of the group are

$$e, a, b, ab,$$

where

$$a^2 = b^2 = (ab)^2 = e.$$

(Verify that these relations imply that the group is Abelian.)

(viii) **Herstein**: *pages 33 and 34, Example 4.*

We shall write $\langle W \rangle$ where Herstein uses (W).

In this example we must, of course, assume that W is non-empty. Then $\langle W \rangle$ contains all products of the form

$$w_{n_1}^{m_1} w_{n_2}^{m_2} \cdots w_{n_k}^{m_k},$$

where each $w_{n_i} \in W$ and the m_i are integers. By the "smallest" subgroup of G containing W, Herstein means that, if H is a subgroup of G such that $W \subset H$, then $\langle W \rangle \subset H$. In *Unit 5, Group Morphisms*, we shall show that $\langle W \rangle$ is a subgroup and that it is the intersection of all subgroups containing W. As mentioned in Section 3.2.2 of *Unit 3*, it follows from the latter assertion that $\langle W \rangle$ is the smallest subgroup containing W.

SAQ 13

Herstein: *page 40, Problem 1.*

(Solution is given on p. 34.)

SAQ 14

G is the group of all integers under addition and H_n is the subgroup consisting of all multiples of a fixed integer n. What is $H_n \cap H_m$?

(Solution is given on p. 34.)

4.4.2   Cosets and Lagrange's Theorem

The binary operation in a group G can be extended in a natural way, so that the product of a subgroup H of G and an element $g \in G$ can be defined. Indeed, let H be a subgroup of a group G and consider the subset of G defined by

$$Hg = \{hg \in G : h \in H\},$$

where g is any particular element of G. That is, for a given $g \in G$, Hg is the set formed by taking the product of all $h \in H$ with g. For demonstration purposes, let $G = S_3$ and let H be the subgroup $\{e, \phi\}$ of S_3. We find that

$$He = H\phi = \{e, \phi\} = H,$$

$$H\psi = H\phi\psi = \{\psi, \phi\psi\},$$

$$H\psi^2 = H\psi\phi = \{\psi^2, \psi\phi\}.$$

In this example, the individual sets obtained have the same number of elements, they are either non-intersecting or identical, and their union is G. The last two properties state that the sets formed by such products constitute a *partition* of G. In this section we shall show that the latter statement is true for any subgroup H of a group G. The analysis of this partition will yield Lagrange's remarkable theorem on finite groups.

The material of this section was introduced in *Unit M100 33, Groups II*. The techniques used here are slightly more general, however. We define an equivalence relation on G, and deduce immediately that the corresponding equivalence classes yield a decomposition of G into disjoint subsets.

READ Herstein: page 34, line 6 to page 37, line 3.

Notes

(i)   *Herstein: page 34, line 9.*
   Instead of Chapter 1 of *Herstein*, refer to *Halmos: Section 7, pages 27–28.*

(ii)   *Herstein: page 34, lines $-2, -1$.*
   We give a direct proof that, if $a \in H$, then $Ha = H$. Let $a \in H$, and h denote a general element of H. Since H is a subgroup, $ha \in H$, i.e.

   $$Ha \subset H.$$

   Also, $h = (ha^{-1})a$, where $ha^{-1} \in H$; hence $h \in Ha$, i.e.

   $$H \subset Ha.$$

   It follows that

   $$Ha = H.$$

   (This result also follows from Herstein's next lemma, Lemma 2.6.)

(iii)   *Herstein: page 35, lines 3 to 6.*
   To make this first paragraph in the proof of Lemma 2.6 a little clearer, we can outline it as follows.

   (1)   We want to show that $ha \in [a]$ for all $h \in H$.
   (2)   We show that $a(ha)^{-1} \in H$.
   (3) $a(ha)^{-1} \in H \Rightarrow a \equiv ha \bmod H$, by the definition of congruence mod H.
   (4)   Thus $ha \in [a]$, by the definition of $[a]$.

(iv) **Herstein**: *page 35, line 18*.

For specific a and b, the mapping is onto, because each element of Hb is of the form hb for some $h \in H$ and, therefore, for any $hb \in Hb$, the element $ha \in Ha$ is mapped to hb.

(v) **Herstein**: *page 35, line -10*.

That is, if the cosets are $H = H_1, H_2, \ldots, H_k$, then $H_i \cap H_j = \varnothing$ for $i \neq j$, $1 \leqslant i, j \leqslant k$.

Notice that no coset mod H other than H is a subgroup, because H is the only coset containing the identity element.

It should be emphasized that there are as many different partitions of G into cosets as there are different subgroups of G. For instance, in S_3, if we consider the subgroup $N = \{e, \psi, \psi^2\}$ instead of H, we obtain the cosets

$$Ne = N\psi = N\psi^2 = \{e, \psi, \psi^2\} = N$$

$$N\phi = N\phi\psi = N\psi\phi = \{\phi, \psi\phi, \phi\psi\}.$$

This is a different decomposition from that obtained from H at the beginning of this section.

(vi) **Herstein**: *page 36, lines 1 and 2*.

This fact will be demonstrated in SAQ 15.

(vii) **Herstein**: *page 37, lines 1–3*.

The group of order 12 which has no subgroup of order 6 is of considerable interest in its own right. In the GROUP CARD package you will find the set of cards for the appropriate group, A_4.

General Comment

Instead of considering Ha, we can develop a similar theory for aH. If H is a subgroup of G and $a \in G$, then the set $aH = \{ah : h \in H\}$ is called a *left coset* of H in G. If we define a new equivalence relation by $a \equiv b$ mod H if $b^{-1}a \in H$, then the theory of left cosets parallels that of right cosets. (We showed this in *Unit M100 36, Mathematical Structures*, Section 36.3.3.) In particular, it follows that the left cosets of H in G decompose G into disjoint subsets and that there is a one-to-one correspondence between any two left cosets. Nevertheless it is easily shown that a left coset aH need not be the same as any right coset of H in G. For example, the left cosets of $H = \{e, \phi\}$ in S_3 are

$$eH = \phi H = \{e, \phi\}$$

$$\psi H = \psi\phi H = \{\psi, \psi\phi\}$$

$$\psi^2 H = \phi\psi H = \{\psi^2, \phi\psi\}$$

and neither of the latter two left cosets is a right coset of H in G.

SAQ 15

Herstein: *page 40, Problem 4*.

(Solution is given on p. 34.)

4.4.3 Consequences of Lagrange's Theorem

Lagrange's Theorem is deceptively easy to interpret and, for this reason, you may not be convinced of the power contained in it. The following implications of the theorem and the exercises at the end of this section should convince you of its importance.

*READ **Herstein**: page 37, line 4 to page 38, the end of Section 4.*

Notes

(i) **Herstein**: page 37, Corollary 1.
You can verify, for example, that each of the elements of S_3 is either of order 2 or of order 3, except for e which, of course, has order 1.

We now have two uses of the word "order":

> $o(G) = $ number of elements in the finite group G;

> $o(a) = $ least positive integer m such that $a^m = e$, where a is an element of a group.

However, there is no possibility of confusion, since Corollary 1 tells us that the order of an element a is the order of the group generated by a, i.e.

> $o(a) = o(\langle a \rangle)$.

(ii) **Herstein**: page 37, line -10 to page 38, line 2, including Corollary 3.
We shall explain this Corollary more explicitly than Herstein. We proved in part (v) of SAQ 7 (see p. 31 of this text) that the set of congruence classes of the integers mod n and relatively prime to n form a group under multiplication mod n. Each such class can be represented by $[a]$ for some $0 < a < n$, where $(a, n) = 1$. It follows immediately that these representative integers (the set of positive integers less than n and relatively prime to n) themselves form a group under multiplication mod n. That is, under multiplication mod n,

> $G = \{a \in J : 0 < a < n \text{ and } (a, n) = 1\}$

is a group, and, by definition of $\phi(n)$,

> $o(G) = \phi(n)$.

Thus, we can apply Corollary 2 to G, obtaining for each $a \in G$ that

> $a^{\phi(n)} = e$.

Now, since $e = 1$ and since the product in G is multiplication mod n, we have

> $a^{\phi(n)} \equiv 1 \bmod n$.

This proves the Corollary for $a \in G$; but we have yet to prove the result for *any* integer a which satisfies $(a, n) = 1$. If we can show that $a \equiv b \bmod n$ for some $b \in G$, we shall have accomplished our purpose, for then

> $a^{\phi(n)} \equiv b^{\phi(n)} \equiv 1 \bmod n$.

So, using the Euclidean algorithm, we write

> $a = kn + b$

(b is the remainder on division by n) for some integer k. Since $(a, n) = 1$,

> $pa + qn = 1$

for some integers p and q. From the last two relations,

> $p(b + kn) + qn = pb + (pk + q)n = 1$

and therefore $(b, n) = 1$. Thus, $a \equiv b \bmod n$, where $b \in G$, and the desired result follows.

(iii) **Herstein**: *page 38, lines 17 to 22.*

Although the expression "genuinely group-theoretic" is a highly subjective one, it is true that the proofs in this section have been much less of a simple grind than those in previous sections. You should attempt to obtain a certain facility with this type of proof before continuing with the text. The following SAQs will help to test whether you have this facility. If necessary, take the statements of the theorems in this section and try to prove them without the book.

SAQ 16

(i)   If G is an infinite group, show that it contains a non-trivial subgroup.
(HINT: For *any* $a \in G$, $a \neq e$, consider the group $\langle a \rangle$.)

(ii) **Herstein**: *page 40, Problem 3.*

(Solution is given on p. 34.)

SAQ 17

Prove that any subgroup of a finite cyclic group is itself a cyclic group.
(HINT: If H is a subgroup of a cyclic group $G = \langle a \rangle$, then each element of H is of the form a^t for some natural number t. Let

$$S = \{t \in \omega : a^t \in H\}.$$

Can you find a generator for H?)

(Solution is given on p. 36.)

SAQ 18

Herstein: *page 41, Problem 16.*

(Solution is given on p. 36.)

SAQ 19

Herstein: *page 41, Problem 15.*
(Suggestion: consider some examples.)

(Solution is given on p. 36.)

4.4.4 A First Counting Principle

A large part of existing group theory is concerned with counting: for example, the order of a group, the number of subgroups in a given group, the number of non-isomorphic groups of order n. Just as Lagrange's Theorem has been shown to provide invaluable assistance in counting processes, so the techniques discussed here provide further hints of what is to come.

In this section, the product HK of two subgroups H and K of G is defined. The condition that HK is itself a subgroup is presented (Lemma 2.8). Then an expression is given for the number of elements in HK, whether or not it is a subgroup (Theorem 2.B). These two results are important and will be used later in this course. The Corollary on page 40 is of only marginal interest, however, and for this reason we suggest that you omit it. In addition, we suggest that you omit the proof of Theorem 2.B.

READ Herstein: page 38, beginning of Section 5 to page 39, line −20, the statement of Theorem 2.B.

Note

(i)   *Herstein: page 39, Theorem 2.B.*
    Recall that $H \cap K$ is a subgroup of G, and thus Lagrange's Theorem implies that

$$o(H \cap K)|o(H) \quad \text{and} \quad o(H \cap K)|o(K).$$

4.5   SUMMARY OF THE TEXT

We have introduced in this unit those properties of the integers which are relevant to the study of group structures in the integers. Also, we have provided a firm foundation for the basic elements in group theory. Our approach has been a more general and more abstract one than you have previously encountered. The concepts and techniques involved in the future study of groups and other algebraic systems will be similar in nature to those which have been developed towards the end of the present sections of *Herstein*.

After the elementary properties of groups have been discovered, the simple and yet important concept of a subgroup opens the door to all the interesting and practical results of group theory. Using the definition of a subgroup H, we develop the equivalence classes of G modulo H (the cosets), and prove Lagrange's important theorem on finite groups.

Lagrange's Theorem is an essential aid in group counting processes. Since it is relevant to almost all results on finite groups, it is worth while to bear it in mind whenever solving a problem.

In this unit, we have defined and used the following concepts:

Integers:

Greatest Common Divisor
Least Common Multiple
Relatively Prime
Prime
Congruence Modulo n
Congruence Classes Modulo n

Groups:

Abelian Group
Order of a Group
Symmetric Group of Degree n
Subgroup
Cyclic Group
Generator of a Cyclic Group
Subgroup Generated by a Set
Order of an Element
Congruence Modulo a Subgroup
Coset
Product of Subgroups

Techniques:

1   Use the Euclidean algorithm to establish relations between given integers.

2   Apply Lemma 2.3 or Lemma 2.4 to determine whether a non-empty subset of a group is a subgroup.

3   In proofs, use the fact that, if a is an element of a group of order n, there must be repetitions in $a^1, \ldots, a^k$ if $k > n$.

4   Use the fact that, in a finite group G, $o(H)|o(G)$ for every subgroup H of G.

5   Use the theorem: if H, K are subgroups of a group G, then

$$HK = KH \Leftrightarrow HK \text{ is a subgroup of } G.$$

Postscript

"The labor union is an elemental response to the human instinct for group action in dealing with group problems."

William Green
(1873–1952)
(Speech, 1925)

4.6 FURTHER SELF-ASSESSMENT QUESTIONS

SAQ 20

Herstein: *page 23, Problem 8.*

(Solution is given on p. 37.)

SAQ 21

Herstein: *page 31, Problem 11.*

(Solution is given on p. 37.)

SAQ 22

Herstein: *page 41, Problem 13.*

(Solution is given on p. 37.)

4.7 SOLUTIONS TO SELF-ASSESSMENT QUESTIONS

Solution to SAQ 1

The steps are as follows:

(i)   Define $\mathfrak{M} = \{x \in J : x = ma + nb, m \in J, n \in J\}$.

(ii) Show that $\mathfrak{M}$ has a least positive integer denoted by

$$c = m_0 a + n_0 b.$$

(iii) Show that $c = (a, b)$ by proving that:

  (1)   any divisor d of a and b is a divisor of c;
  (2) c is a divisor of a and b.

Your outline should have at least as much detail as the above, although it could of course have more. (If we wish to find values for m and n, we can use the Euclidean algorithm, which we described in *Unit M100 34, Number Systems*, Section 34.2.4.)

Solution to SAQ 2

For the existence of d, we follow an outline as in SAQ 1, but now we include the details of proof.

(i)   For given non-zero integers a and b, define

$$S = \{s \in J : s > 0, a|s \text{ and } b|s\}.$$

(ii) If $ab > 0$, then $ab \in S$. If $ab < 0$ then $-ab \in S$. Thus $S \neq \varnothing$ and S has a least element g.

(iii) We now show that g is the least common multiple of a and b by showing that

  (1) $a|g$ and $b|g$;
  (2)   if $a|x$ and $b|x$, then $g|x$.

  Firstly, because $g \in S$, we know that $g > 0$, $a|g$ and $b|g$. Thus g satisfies the first requirement. Secondly, suppose that $a|x$ and $b|x$. If $x = 0$, then $g|x$. If $x \neq 0$, then $|x| \in S$. Thus $|x| \geqslant g$ and, by the Euclidean algorithm,

  $$|x| = mg + r,$$

  for some integers m and r, $0 \leqslant r < g$. Now, since $a\,|\,|x|$ and $a|g$, it follows from this equation that $a|r$ (see Note (iii) on page 7 of this unit). Similarly, $b|r$. Consequently, $r = 0$, for otherwise $r \in S$ and $r < g$, which contradicts the definition of g. Therefore

  $$|x| = mg,$$

  so that $g\,|\,|x|$ and $g|x$. We have thus verified the second requirement, proving that g is the least common multiple of a and b.

  To prove uniqueness, assume that d_1 and d_2 both satisfy the two properties of the least common multiple of a and b. Then, using property (1) for d_2 and property (2) for d_1, we obtain

  $$(a|d_2 \quad \text{and} \quad b|d_2) \Rightarrow d_1|d_2.$$

  Similarly, using property (1) for d_1 and property (2) for d_2, we have $d_2|d_1$. Lastly, since both d_1 and d_2 are positive, we must have $d_1 = d_2$.

Solution to SAQ 3

Suppose first that $p|a_1 a_2$, where p is a prime. If p is relatively prime to a_1, then by Lemma 1.6, $p|a_2$. If p is not relatively prime to a_1, then

$$(p, a_1) = c > 1.$$

But since $c|p$ and p is prime, we must have $c = p$. Thus, because $c|a_1$, it follows that $p|a_1$. This completes the proof for a product of two integers.

Proceeding by induction, assume that

$$p|a_1 a_2 \cdots a_{n-1} \Rightarrow p|a_i \qquad \text{for some} \quad i = 1, 2, \ldots, n-1,$$

and suppose that

$$p|a_1 a_2 \cdots a_{n-1} a_n.$$

If we put $b = a_1 a_2 \cdots a_{n-1}$, then this expression becomes $p|ba_n$. From the first step of this proof it follows that

$$p|b \qquad \text{or} \qquad p|a_n.$$

From our assumption, we have therefore that

$$p|a_i \qquad \text{for some} \quad i = 1, 2, \ldots, n-1 \quad \text{or} \quad p|a_n.$$

This is the desired assertion.

Solution to SAQ 4

We must show that for $n > 1$:

(i) if n is prime, then for any a, either $(a, n) = 1$ or $n|a$;
(ii) if, for any a, either $(a, n) = 1$ or $n|a$, then n is prime.

The proofs are as follows.

(i) If n is prime, then by definition, its only divisors are 1 and n. But since $(a, n)|n$, this means that either $(a, n) = 1$ or $(a, n) = n$. In the latter case, it follows from $(a, n)|a$ *that* $n|a$.

(ii) Suppose that, for all $a \in J$, *either* $(a, n) = 1$ *or* $n|a$.

Let d be a positive integer such that $d|n$.

Then $(d, n) = d$, so *either* $d = 1$ *or* $n|d$. But

$$(d|n \text{ and } n|d) \Rightarrow d = n,$$

so *either* $d = 1$ *or* $d = n$. It follows that n is a prime number.

Solution to SAQ 5

By Lemma 1.5, there exist integers b and k such that

$$ab + nk = 1.$$

Therefore

so

$$ab \equiv 1 \bmod n \qquad \text{(definition of congruence mod } n\text{)}$$

$$[ab] = [1] \qquad \text{(definition of congruence classes)}$$

and

$$[a][b] = [1] \qquad \text{(definition of product for congruence classes)}.$$

Solution to SAQ 6

(i) We show first that, if n is a prime number, then

$$[a][b] = [0] \Rightarrow ([a] = [0] \qquad \text{or} \qquad [b] = [0]).$$

Since

$$[a][b] = [ab],$$

$[a][b] = [0]$ implies that $n|ab$. Therefore, by the Corollary in **Herstein**, *page 19*, $n|a$ or $n|b$, so that by the definition of congruence mod n,

$$[a] = [0] \qquad \text{or} \qquad [b] = [0].$$

(ii) Conversely, we show that, if

$$[a][b] = [0] \Rightarrow ([a] = [0] \qquad \text{or} \qquad [b] = [0]),$$

then n is a prime.

Let a and b be positive integers such that $ab = n$.

Then $[a][b] = [0]$, and so $[a] = [0]$ or $[b] = [0]$.

If $[a] = [0]$, then $n|a$; since $a|n$, it follows that $a = n$ and $b = 1$. Similarly, if $[b] = [0]$, then $b = n$ and $a = 1$. Hence n is prime.

Solution to SAQ 7

(i) This is a group. Closure follows from the expression

$$\frac{p_1}{q_1} \times \frac{p_2}{q_2} = \frac{p_1 p_2}{q_1 q_2} = \frac{p}{q},$$

and associativity follows from

$$\frac{p_1}{q_1} \times \left(\frac{p_2}{q_2} \times \frac{p_3}{q_3}\right) = \frac{p_1 p_2 p_3}{q_1 q_2 q_3} = \left(\frac{p_1}{q_1} \times \frac{p_2}{q_2}\right) \times \frac{p_3}{q_3}.$$

The identity element is $1 = 1/1$, and the inverse of p/q is q/p.

(ii) This is not a group. The first three axioms are satisfied and 1 is the identity. However, 0 has no inverse because

$$0 \cdot s = 0 \neq 1$$

for all real s. If we exclude 0 as in (i), the system becomes a group.

(iii) Closure follows from the definition of addition of congruence classes mod n (**Herstein**: *page 22, Equation (a)*). Associativity was stated in **Herstein**: *page 22, Equation (3)*. The fact that $[0]$ is the identity follows from **Herstein**: *page 22, Equations (1) and (6)*:

$$[0] + [i] = [i] + [0] = [i].$$

For any $[i]$, we have

$$[i] + [n - i] = [i + n - i] = [0].$$

Therefore, the inverse of any element $[i]$ is $[n - i]$ which is also an element of J_n. J_n is therefore a group under addition.

(iv) We saw in SAQ 6 that, if $n > 1$ is not prime, there exist $[a], [b] \in J_n - \{[0]\}$ such that $[a][b] = [0]$.

Thus, $(J_n - \{[0]\}$, multiplication) will not be closed and thus will not be a group if n is not prime.

However, if $n > 1$ is prime, SAQ 6 tells us that, for all elements $[a], [b] \in J_n - \{[0]\}$,

$$[a][b] \neq [0].$$

Thus, closure is satisfied if n is prime. Associativity and the verification that $[1]$ is the identity follow from **Herstein**: *page 22, Equations (4) and (7)*. Lastly, if n is prime, $(a, n) = 1$ for all $a \neq 0 \bmod n$, and so SAQ 5 tells us that each $a \in J_n - \{[0]\}$ has an inverse under multiplication.

Thus, if n is prime, $(J_n - \{[0]\}$, multiplication mod n) is a group.

In the case $n = 1$, $(J_n - \{[0]\}$, multiplication) is a trivial group with one element, as is easily verified.

(v) To verify that $(G, \circ)$ is closed, we proceed as above; we must also show that

$$((a, n) = (b, n) = 1) \Rightarrow (ab, n) = 1.$$

From SAQ 5, we know that there exist integers k and l such that $[k][a] = [1]$ and $[l][b] = [1]$. Thus

$$[klab] = ([k][a])([l][b]) = [1],$$

and therefore,

$$(kl)ab + tn = 1$$

for some integer t. It follows that

$$(ab, n) = 1.$$

The proofs of associativity, identity and inverses are as in (iv).

Solution to SAQ 8

(i)

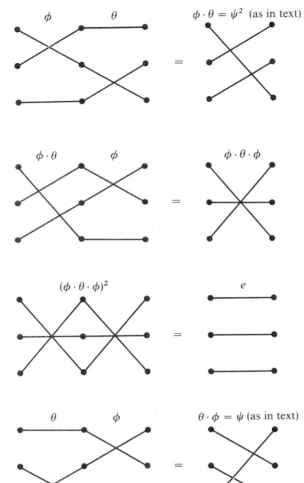

So $(\theta \cdot \phi)^2 = \psi^2 = \phi \cdot \theta$ by part (i).

(iii) From the above and our results on page 16, we know that

$$\theta \cdot \phi = \psi \qquad \text{and} \qquad \phi^2 = e,$$

so

$$\theta = \theta \cdot \phi \cdot \phi = \psi \cdot \phi.$$

We know that

$$S_3 = \{e, \phi, \psi, \psi^2, \phi \cdot \psi, \psi \cdot \phi\}$$

(under composition), so that

$$S_3 = \{e, \phi, \theta \cdot \phi, (\theta \cdot \phi)^2 = \phi \cdot \theta, \phi \cdot \theta \cdot \phi, \theta \cdot \phi \cdot \phi = \theta\}$$

and

$$\phi \cdot \theta \cdot \phi = (\phi \cdot \theta) \cdot \phi = (\theta \cdot \phi \cdot \theta \cdot \phi) \cdot \phi = \theta \cdot \phi \cdot \theta.$$

Therefore,

$$S_3 = \{e, \phi, \theta \cdot \phi, \phi \cdot \theta, \theta \cdot \phi \cdot \theta, \theta\}.$$

Solution to SAQ 9

(i) $(J_n$, addition mod $n)$.

(ii) For $n + 1$ prime, $(J_{n+1} - \{[0]\}$, multiplication mod $n + 1)$.

Solution to SAQ 10

Consider first $n \geqslant 0$. For $n = 0$ and $n = 1$, the assertion is trivial. Assume the assertion to be true for $0, 1, \ldots, n - 1$. Then, since G is Abelian,

$$
\begin{aligned}
(a \cdot b)^n &= (a \cdot b) \cdot (a \cdot b)^{n-1} && \text{(definition of } (a \cdot b)^n) \\
&= (a \cdot b) \cdot (a^{n-1} \cdot b^{n-1}) && \text{(induction hypothesis)} \\
&= (a \cdot b) \cdot (b^{n-1} \cdot a^{n-1}) && \text{(commutativity)} \\
&= a \cdot (b \cdot (b^{n-1} \cdot a^{n-1})) && \text{(associativity)} \\
&= a \cdot ((b \cdot b^{n-1}) \cdot a^{n-1}) && \text{(associativity)} \\
&= a \cdot (b^n \cdot a^{n-1}) && \text{(definition of } b^n) \\
&= a \cdot (a^{n-1} \cdot b^n) && \text{(commutativity)} \\
&= (a \cdot a^{n-1}) \cdot b^n && \text{(associativity)} \\
&= a^n \cdot b^n && \text{(definition of } a^n).
\end{aligned}
$$

Next, consider $n < 0$. Let $p = -n$. Then p is positive and we have

$$
\begin{aligned}
(a \cdot b)^n &= ((a \cdot b)^{-1})^p \\
&= (b^{-1} \cdot a^{-1})^p && \text{(inverse of } a \cdot b) \\
&= (a^{-1} \cdot b^{-1})^p && \text{(commutativity)} \\
&= (a^{-1})^p \cdot (b^{-1})^p && \text{(by first part of proof)} \\
&= a^{-p} \cdot b^{-p} && \text{(by \textbf{Herstein}: page 28, Equation (2))} \\
&= a^n \cdot b^n.
\end{aligned}
$$

Solution to SAQ 11

Suppose that G is Abelian. Then

$$(a \cdot b)^2 = (a \cdot b) \cdot (a \cdot b) = a \cdot (b \cdot a) \cdot b = a \cdot (a \cdot b) \cdot b = a^2 \cdot b^2.$$

The converse is proved as follows.

For any $a, b \in G$,

$$(a \cdot b)^2 = (a \cdot b) \cdot (a \cdot b) = a \cdot b \cdot a \cdot b$$

and

$$a^2 \cdot b^2 = a \cdot a \cdot b \cdot b.$$

Thus, by hypothesis,

$$a \cdot b \cdot a \cdot b = a \cdot a \cdot b \cdot b$$

so that, using left cancellation of a and right cancellation of b, we have

$$b \cdot a = a \cdot b$$

for any $a, b \in G$, and so G is Abelian.

Solution to SAQ 12

Let $a \in G$, where $o(G) = n$, and consider the $n + 1$ elements $a^0, a^1, \ldots, a^n$. Each of these powers is an element of G. But since G has only n elements, at least two of these powers must be equal; that is, for some i and j, $0 \leqslant i < j \leqslant n$,

$$a^j = a^i.$$

Therefore,

$$a^j a^{-i} = a^i a^{-i}$$

and

$$a^{j-i} = e.$$

If we denote the elements of G by $a_1, \ldots, a_n$, we have proved that, for each a_m, there exists a positive integer k_m such that

$$a_m^{k_m} = e.$$

Let $N = k_1 k_2 \cdots k_n$. For the element $a_m \in G$, let $k = N/k_m$. Then

$$a_m^N = a_m^{k_1 k_2 \cdots k_n} = (a_m^{k_m})^k = e^k = e,$$

and the proof is complete. Note that we could take N to be any integer such that $k_m | N$ for all $m = 1, 2, \ldots, n$.

Solution to SAQ 13

We shall use **Herstein**: *page 32, Lemma 2.3.* (Since $e \in H, e \in K, H \cap K$ is non-empty.)

(1) $a, b \in H \cap K \Rightarrow a, b \in H$ and $a, b \in K$

$\Rightarrow ab \in H$ and $ab \in K$ since H and K are subgroups

$\Rightarrow ab \in H \cap K$.

(2) $a \in H \cap K \Rightarrow a \in H$ and $a \in K$

$\Rightarrow a^{-1} \in H$ and $a^{-1} \in K$

$\Rightarrow a^{-1} \in H \cap K$.

Hence $H \cap K$ is a subgroup of G.

Solution to SAQ 14

We show that

$$H_n \cap H_m = H_d,$$

where d is the least common multiple of n and m. (See SAQ 2, page 8.)

$g \in H_d \Leftrightarrow g = kd$ for some integer k

$\Leftrightarrow n|g$ and $m|g$, by the definition of least common multiple

$\Leftrightarrow g = pn$ and $g = qm$ for integers p, q

$\Leftrightarrow g \in H_n$ and $g \in H_m$

$\Leftrightarrow g \in H_n \cap H_m$.

Solution to SAQ 15

We have $H_n = \{a \in G : a = pn \text{ for some } p \in G\}$.

Let $i \in G$. The group operation is addition, so it seems more appropriate to denote a right coset of H_n by $H_n + i$ rather than $H_n i$. Thus

$$H_n + i = \{a \in G : a = pn + i \text{ for some } p \in G\},$$

and thus $H_n + i = [i]$, the congruence class of i, modulo n. Therefore,

$$H_n + i = H_n + j$$

if and only if $i \equiv j \bmod n$. The cosets of H_n in G are

$$H_n = [0], \qquad H_n + 1 = [1], \qquad H_n + 2 = [2], \ldots, \qquad H_n + (n-1) = [n-1],$$

and so the index of H_n in G is n.

Solution to SAQ 16

(i) For any $a \in G$, $a \neq e$, $\langle a \rangle$ is either finite or infinite. If $\langle a \rangle$ is finite, it is the desired subgroup. If it is infinite, it is either a non-trivial subgroup or $\langle a \rangle = G$. If $\langle a \rangle = G$, consider the group $H = \langle a^2 \rangle$. $a \notin H$, because if $a \in H$, there would exist an i such that $a^{2i} = a$, so $a^{2i-1} = e$, contradicting the assumption that $\langle a \rangle$ was infinite. Thus, if $G = \langle a \rangle$, $\langle a^2 \rangle$ is a non-trivial subgroup and the proof is complete.

The structure of this proof can be expressed by the following flow chart, in which H represents a non-trivial subgroup.

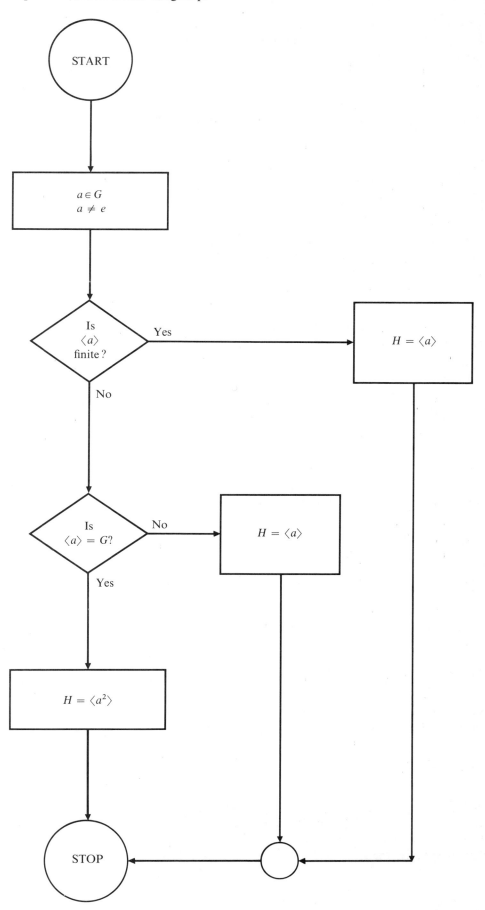

(ii)   Because of (i), G must be finite. For any $a \in G, a \neq e$, the order of a must be $n = o(G)$, for otherwise G would have a non-trivial subgroup. Now, suppose $n = pq$ for some positive integers p and q both greater than 1. Then $a^{pq} = e$ or $(a^p)^q = e$. But that would mean that the element $b = a^p$ has order $q < n$, contradicting the hypothesis that G has no non-trivial subgroups. Thus $n \neq pq$ for any integers p and q, both greater than one, so n must be prime.

Solution to SAQ 17

Since $S - \{0\}$ is a set of positive integers, it has a least element k. Let a^t be any element in H, where $t \neq k$. Then

$$t = km + l,$$

for some integers m and l, where $0 \leqslant l < k$. Since $a^k \in H$, we have that a^{km} and thus a^{-km} are also in H. Therefore

$$a^l = a^{t-km} = a^t a^{-km} \in H.$$

This means that $l = 0$, for otherwise $l \in S$ and $l < k$, which is a contradiction. Therefore $k|t$ and so $a^t \in \langle a^k \rangle$. Since a^t was an arbitrary element of H, it follows that $H \subset \langle a^k \rangle$. However, by closure, $\langle a^k \rangle \subset H$. Thus $H = \langle a^k \rangle$.

Solution to SAQ 18

Suppose, to the contrary, that $o(a) \nmid m$. In other words, suppose that

$$m = ko(a) + s, \text{ where } 0 < s < o(a).$$

Then

$$e = a^m = a^{ko(a)+s} = (a^{o(a)})^k \cdot a^s = a^s.$$

But since $0 < s < o(a)$, $a^s = e$ contradicts the definition of $o(a)$.

Solution to SAQ 19

Let $G = \langle a \rangle$ and $n = o(a) = o(G)$. If $b \in G$, then $b = a^k$ for some $k < n$.

Suppose $(k, n) = s \neq 1$, and let $n = ps, k = qs$, where $0 < p < n, 0 < q < k$. Then by Corollary 2, **Herstein**, *page 37*,

$$(a^k)^p = (a^{qs})^p = (a^q)^n = e.$$

But since $p < n, b = a^k$ cannot be a generator of G.

Now suppose $(k, n) = s = 1$. Since

$$b^{o(b)} = (a^k)^{o(b)} = e,$$

we have from SAQ 18 that $n|ko(b)$. But since $(k, n) = 1$, this means that $n|o(b)$ (**Herstein**, *page 19*, Lemma 1.6). However, from Corollary 1, **Herstein**, *page 37*, we have that $o(b)|n$. Thus $n = o(b)$ and b is a generator of G.

Therefore, the number of generators of a cyclic group of order n is equal to the number of integers less than n and relatively prime to n. But that is just the definition of the Euler function $\phi(n)$.

Solution to SAQ 20

Suppose that n is not divisible by any prime $p \leqslant \sqrt{n}$ and suppose that $n = ab$ for some integers a and b, both greater than 1. Since a itself is a product of primes by the unique factorization theorem, and since any such prime is also a divisor of n, we would have $a > \sqrt{n}$. Similarly, under the given hypothesis, $b \quad \sqrt{n}$. But then

$$n = ab > \sqrt{n}\sqrt{n} = n,$$

which contradicts the statement that $n = ab$. Thus n must be prime.

Solution to SAQ 21

Consider all distinct subsets (unordered pairs) of G of the form

$$\{a, a^{-1}\}.$$

Since each element of G has a unique inverse, each element is contained in some such subset and no element is contained in two different subsets. (These subsets form a partition of G.) $\{e\}$ is such a subset. Let k be the total number (finite!) of distinct subsets. If all subsets besides $\{e\}$ contain two elements, then

$$o(G) = 2k - 1,$$

which is a contradiction. Thus, at least one subset $\{b\}, b \neq e$, contains only one element; whence $b = b^{-1}$.

Solution to SAQ 22

The *centre* of G is the set of all elements of G which commute with every element of G. Notice that the centre is necessarily Abelian.

To prove that the centre Z of G is a subgroup of G, suppose $z_1, z_2 \in Z$. Then for any element $x \in G$,

$$(z_1 z_2)x = z_1(z_2 x) = z_1(x z_2) = (z_1 x)z_2 = (x z_1)z_2 = x(z_1 z_2),$$

so that $z_1 z_2 \in Z$.

Now consider $z^{-1} \in Z$ and let x be any element of G. Then

$$x = ex = (z^{-1}z)x = z^{-1}(xz) = (z^{-1}x)z,$$

so that, taking the product by z^{-1} on the right,

$$xz^{-1} = z^{-1}x,$$

so

$$z \in Z \Rightarrow z^{-1} \in Z.$$

Consequently, by Lemma 2.3, Z is a subgroup of G.

Unit 5   Group Morphisms

"Now this establishment of correspondence between two aggregates and investigation of the propositions that are carried over by the correspondence may be called the central idea of modern mathematics."

W. K. Clifford
Philosophy of the Pure Sciences;
Lectures and Essays (London, 1901),
Vol. 1, p. 402.

Contents

Set Books

P. R. Halmos, *Naive Set Theory*, paperback edition 1972 (Van Nostrand Reinhold).
I. N. Herstein, *Topics in Algebra*, paperback edition (Xerox College/T.A.B.S, 1964).
B. Mendelson, *Introduction to Topology*, paperback edition 1972 (Allyn and Bacon).
M. L. Minsky, *Computation: Finite and Infinite Machines*, paperback edition 1972 (Prentice-Hall).

It is essential to have these books; the course is based on them and will not make sense without them.

Unit 5 is based on **Herstein**, Chapter 2, Sections 6 and 7.

Conventions

Before working through this correspondence text make sure you have read *A Guide to the Course: Topics in Pure Mathematics*.

References to the Open University Mathematics Foundation Course Units (The Open University Press, 1971) take the form *Unit M100 3, Operations and Morphisms*.

5.0 INTRODUCTION

In *Unit 4, Group Axioms*, we discussed the axioms for a group and the concept of a *subgroup*, and we established some simple lemmas and theorems on the existence of subgroups of a given group.

The group axioms are satisfied by many mathematical systems, so it is useful to examine their consequences. A glance at any book on group theory should convince you that the four group axioms have very many consequences, and so it is necessary to have some idea of where we are going and what we want to prove.

There are two main approaches. First, we want to learn how to break down a group into constituent groups. Secondly, we want to learn how to put the constituents back together—in other words, to build up new groups from old. In both cases we want to know how group-theoretic properties transfer from a group to its constituents, and from the constituents to a constructed group. The key to these problems is provided by group morphisms.

The constituents that are always useful are kernels of morphisms and homomorphic images. Sometimes the straightforward substructure (subgroups) can be useful as well.

You first met these ideas in *Unit M100 23, Linear Algebra II*, where we discussed linear transformations of vector spaces, and the notion of a kernel. In *Unit M100 33, Groups II*, we applied the same ideas to groups. We introduced the idea of a factor group or quotient group (the names are used interchangeably) by looking at group morphisms, and we again discovered the importance of the kernel. We then demonstrated the First Isomorphism Theorem which states that every image group under a morphism is isomorphic to a quotient group. In this unit we shall review these ideas, and extend them further.

This unit is about group morphisms or, equivalently, about quotient groups, since a quotient group is the image of a group under a morphism. We introduce the concept of a *normal subgroup* of a group and show that the kernel of a morphism of a group G is a normal subgroup of G, and, conversely, any normal subgroup of a group G can be used to define a morphism from G to a quotient group. This important result is the heart of group theory. The value of the kernel and the quotient structure is that together they carry a good deal of the information stored in the group. The kernel carries details, and the quotient structure carries an overall view in which some detail has been obscured.

In this unit, we shall go much further than in the Foundation Course, by applying the First Isomorphism Theorem to prove two theorems for finite Abelian groups which are special cases of extremely important group theory results. These theorems tell us about the existence of certain subgroups of a group G. The remainder of the unit is taken up with discussions of the subgroups and quotient groups of a quotient group. The last section discusses the most important quotient group, involving the commutator subgroup.

Aims

After working through this text, you should be able to:

(i)   construct homomorphisms and isomorphisms between appropriate groups;

(ii)   perform computations involving group elements and sets of group elements;

(iii)   demonstrate and use the basic properties of homomorphisms, and, in particular, the First, Second and Third Isomorphism Theorems;

(iv)   demonstrate the basic properties of normal subgroups such as:

> if N is a normal subgroup of G, and H is a subgroup, then NH is a subgroup of G, and $N \cap H$ is normal in H;

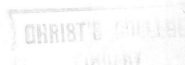

5

(v) demonstrate the basic properties of the commutator subgroup G' of a group G, such as:

> G' is normal in G,
> G/G' is Abelian,
> G/N is Abelian implies $G' \subset N$.

You should also be aware of the strategies used to prove group theory theorems, as shown in this unit.

5.1   NORMAL SUBGROUPS

5.1.0   Introduction

In this section we discuss the extremely important concept of a *normal subgroup*. In *Unit 4, Group Axioms*, we used the right coset equivalence classes of G modulo an arbitrary subgroup H to prove Lagrange's Theorem. Although any subgroup of G can be used to split G into right cosets, only the normal subgroups permit the set of right cosets to have an induced group structure. In other words, the group multiplication is compatible with the equivalence relation (see *Unit M100 19, Relations*) only when the relation comes from the right cosets of a *normal* subgroup.

5.1.1   Definition of a Normal Subgroup

*READ **Herstein**: page 41, beginning at Section 6, to page 43, line -14 (Proof of Lemma 2.11).*

Notes

(i)   ***Herstein**: page 42, lines 18–21 (Definition).*
This definition is very important. Herstein defines a subgroup N of a group G to be a normal subgroup of G if

for *all* $g \in G$, and for all $n \in N$, $gng^{-1} \in N$.

Denoting $\{gng^{-1} : n \in N\}$ by gNg^{-1}, we may write the above condition in the form

for *all* $g \in G$, $gNg^{-1} \subset N$.

The words *for all* mean that we may replace g by g^{-1}, and use the fact that $(g^{-1})^{-1} = g$, to obtain the equivalent condition

for all $g \in G$, $g^{-1}Ng \subset N$.

In Lemma 2.9, Herstein shows that the conditions

$$\forall g \in G \quad gNg^{-1} \subset N$$

$$\forall g \in G \quad g^{-1}Ng \subset N$$

$$\forall g \in G \quad gNg^{-1} = N$$

(and hence

$$\forall g \in G \quad g^{-1}Ng = N)$$

are equivalent.

(ii)   *Examples of Normal Subgroups*

(a)   Every subgroup of an Abelian group is normal. (See SAQ 1, page 8.)
(b) S_3. The subgroup $N = \{e, \psi, \psi^2\}$ is normal in S_3. To see this, place the group cards in a column as shown.

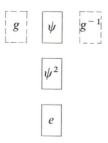

For the remaining cards, observe that each is its own inverse (each has order 2). Calculate gng^{-1} for each $n \in N$ and each $g \notin N$, and observe that in each case it is a member of N. In this manner, each of the elements g is permuting the elements of the set N.

(c) A_4. The subgroup $N = \{e, a, b, ab\}$ is normal in A_4. Check this in the same way as for S_3, but observe that the elements $g \notin N$ have order 3.

(iii) **Herstein**: *page 43, lines 1 to 3* (*Proof of Lemma 2.10*).
In symbols, we have:
$$N \text{ normal} \Rightarrow gNg^{-1} = N \qquad \text{for all} \quad g \in G \text{ (Lemma 2.9)}$$
$$\Rightarrow (gNg^{-1})g = Ng$$
$$\Rightarrow gN(g^{-1}g) = Ng$$
$$\Rightarrow gN = Ng.$$

(iv) **Herstein**: *page 43, line 10.*
The fact that two distinct right cosets have no element in common was used in the proof of Lagrange's Theorem, **Herstein**: *page 35*. It is a consequence of the fact that distinct right cosets are distinct equivalence classes.

(v) **Herstein**: *page 43, line -14* (*Proof of Lemma 2.11*).
The unproved half is SAQ 3.

Example

Show that, in A_4, the subgroup $H = \{e, c, c^2\}$ is not normal, by finding a left coset which is not a right coset.

Solution

Look at the left coset aH. By direct calculation with the group cards, $aH = \{a, ac, ac^2\}$.

The right coset Ha is found to be $\{a, bc, abc^2\}$.

In this group, the only right coset of H which is also a left coset is the coset H itself.

General Comment

Let H be a subgroup of a group G. Later we shall find it convenient to define a mapping T_g associated with the element $g \in G$ as follows:
$$T_g : H \longrightarrow g^{-1}Hg,$$
where
$$T_g : h \longmapsto g^{-1}hg \qquad (h \in H).$$
Writing mappings on the right, we have
$$HT_g = g^{-1}Hg, \qquad HT_{g^{-1}} = gHg^{-1},$$
and
$$hT_g = g^{-1}hg, \qquad hT_{g^{-1}} = ghg^{-1}.$$
A subgroup H of G is normal if and only if
$$\text{for all } g \in G, \qquad HT_g \subset H$$
or, equivalently,
$$\text{for all } g \in G, \qquad HT_{g^{-1}} \subset H.$$

SAQ 1

Herstein: *page 45, Problem 6.*

(Solution is given on p. 37.)

8

SAQ 2

Herstein: *page 45, Problems 8 and 9.*
(HINT: In Problem 9, use the result of Problem 8.)

(Solution is given on p. 37.)

SAQ 3
Herstein: *page 44, Problem 1.*
(This is the completion of Lemma 2.11.)

(Solution is given on p. 37.)

5.1.2 Quotient Groups

READ Herstein: page 43, line −13 to page 44, the end of the section.

Note

(i) **Herstein**: *page 43, lines −11 to −5.*
Having defined earlier what we mean by the product of two subsets of a group G, we are going to use this product as a binary operation on G/N. The elements of this new structure are now cosets of N in G.

The formula

$$NaNb = Nab$$

tells us that the following diagrams are commutative:

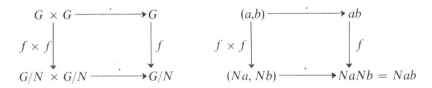

(Since f maps G to G/N, we use $f \times f$ to represent the obvious map from $G \times G$ to $G/N \times G/N$.)

So the mapping f, which maps $a \in G$ to $Na \in G/N$, preserves the group operation. The next part of the present reading passage proves that G/N, together with this binary operation, has in fact inherited all the group properties and is itself a group.

General Comment

Frequently, in the next few sections, we shall be given a homomorphism from a group G to a group H,

$$\phi : G \longrightarrow H,$$

and wish to construct a function ψ with domain a quotient group G/N and codomain H. Since the quotient group is a set of cosets of the form Ng, the obvious specification of ψ is

$$\psi : Ng \longrightarrow \phi(g),$$

because then the diagram

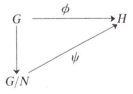

will commute. Great care must be taken, however, to ensure that the ψ specified as above is in fact a function. Since various representatives can be chosen from the same coset, we must first check that

$$Ng = Nh \Rightarrow \psi(Ng) = \psi(Nh),$$

and this requires that

$$\phi(g) = \phi(h).$$

This check is referred to as checking that ψ is *well-defined*. (We met this term in *Unit 4, Group Axioms*, when we discussed operations on the equivalence classes of the integers modulo n, a situation completely analogous to the present one. See *Herstein: page 22, line 9*.)

In a more general framework, suppose we are given a function ϕ from a set S to a set T, and another function f on S. Then we wish to complete the following diagram and make it commutative:

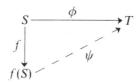

Clearly, we want to choose

$$\psi(f(x)) = \phi(x),$$

for each $x \in S$. The mapping ψ so defined will be a function if

$$f(x) = f(y) \Rightarrow \phi(x) = \phi(y).$$

So, in this general situation, we say that ψ is *well-defined* if the latter condition holds.

Example

In the group A_4, let $N = \{e, a, b, ab\}$, which we showed to be normal in Example (c) of Section 5.1.1. Find the order of A_4/N. Describe the group A_4/N. (You may find the group cards helpful.)

Solution

Since $o(A_4) = 12$ and $o(N) = 4$, it follows that $o(A_4/N) = 3$. This conforms with the fact that there are three right cosets of N in A_4:

$$N = \{e, a, b, ab\},$$

$$Nc = \{c, ac, bc, abc\},$$

$$Nc^2 = \{c^2, ac^2, bc^2, abc^2\}.$$

A_4/N is a cyclic group of order 3.

SAQ 4

Herstein: *page 45, Problem 2.*

(Solution is given on p. 38.)

SAQ 5

(i) **Herstein**: *page 45, Problem 3.*
     (Show that $HN = NH$.)

(ii) If N is a subgroup of H, H is a subgroup of G and N is normal in G, show
      that H/N is a subgroup of G/N.

$$
\begin{array}{ccl}
G & \longrightarrow & G/N \\
| & & | \\
H & \longrightarrow & H/N \\
| & & | \\
N & \longrightarrow & N/N = \langle \bar{e} \rangle
\end{array}
$$

(Solution is given on p. 38.)

5.2 HOMOMORPHISMS

5.2.0 Introduction

This section reviews Foundation Course material. There, you became familiar with the idea of a morphism. Herstein uses homomorphism to mean morphism, and so, unlike the Foundation Course, permits his homomorphisms to be one-to-one. Herstein assumes no knowledge of morphisms. He introduces the idea, and then proceeds to link it to the idea of kernel (which for a group is synonymous with a normal subgroup). This investigation will culminate in Section 5.3.1 with the First Isomorphism Theorem, which you also met in the Foundation Course.

Notice that Herstein writes mappings on the right when he talks about a mapping from a set into itself, but otherwise he usually writes mappings *on the left*.

5.2.1 Definitions and Examples

*READ **Herstein**: page 46, the beginning of Section 7, to page 48, line −1, the end of Lemma 2.16.*

Notes

(i) *Herstein: page 46, line 13.*
Notice that homomorphisms are permitted to be one-to-one, and not necessarily onto. Herstein's usage is the more common.

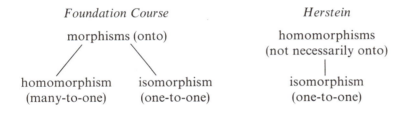

(ii) *Herstein: page 46, line − 18 (Example 0).*
Notice that, if G and H are groups, and $h_0 \neq e$, $h_0 \in H$, then the constant function

$$\phi : g \longmapsto h_0 \qquad (g \in G)$$

is a function but *not* a homomorphism; for if $g_1, g_2 \in G$, then

$$\phi(g_1 g_2) = h_0$$

and

$$\phi(g_1)\phi(g_2) = h_0^2.$$

But $h_0 = h_0^2$ implies that $h_0 = e$, which is a contradiction.

(iii) *Herstein: page 46, line −8 (Example 1).*
$G = (R, +)$ and $\bar{G} = (R − \{0\}, \times)$

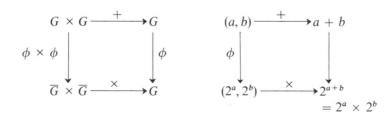

(iv) **Herstein**: *page 46, line −4 (Example 2).*
The action of f on the group table for S_3 can be displayed as follows:

S_3	e	ψ	ψ^2	ϕ	$\phi\psi$	$\phi\psi^2$
e	e	ψ	ψ^2	ϕ	$\phi\psi$	$\phi\psi^2$
ψ	ψ	ψ^2	e	$\phi\psi^2$	ϕ	$\phi\psi$
ψ^2	ψ^2	e	ψ	$\phi\psi$	$\phi\psi^2$	ϕ
ϕ	ϕ	$\phi\psi$	$\phi\psi^2$	e	ψ	ψ^2
$\phi\psi$	$\phi\psi$	$\phi\psi^2$	ϕ	ψ^2	e	ψ
$\phi\psi^2$	$\phi\psi^2$	ϕ	$\phi\psi$	ψ	ψ^2	e

$\xrightarrow{\quad f \quad}$

$\overline{G}$	e	e	e	ϕ	ϕ	ϕ
e	e	e	e	ϕ	ϕ	ϕ
e	e	e	e	ϕ	ϕ	ϕ
e	e	e	e	ϕ	ϕ	ϕ
ϕ	ϕ	ϕ	ϕ	e	e	e
ϕ	ϕ	ϕ	ϕ	e	e	e
ϕ	ϕ	ϕ	ϕ	e	e	e

When redundancies are removed, the group table for $\overline{G}$ becomes:

$\overline{G}$	e	ϕ
e	e	ϕ
ϕ	ϕ	e

(v) **Herstein**: *page 47, line −2.*
Herstein occasionally uses "unit element" as an alternative term for the identity element in a group. To show that $\phi(e)$ is the identity $\bar{e}$ in $\overline{G}$, we try to arrange matters so that we can "pull back" the calculation into the group G. Thus for any $x \in G$, $\phi(x) \in \overline{G}$, and so

$$\phi(x)\bar{e} = \phi(x) \text{ in } \overline{G}.$$

But

$$x = xe \text{ in } G,$$

so

$$\phi(x) = \phi(xe)$$
$$= \phi(x)\phi(e),$$

since ϕ is a homomorphism.

Thus

$$\phi(x)\bar{e} = \phi(x)\phi(e),$$

whence

$$\bar{e} = \phi(e) \qquad \text{(cancellation property in } \overline{G}).$$

(vi) **Herstein**: *page 48, line −17.*
The previous arguments permitted ϕ to be *into*, but now it must be *onto*.

(vii) **Herstein**: *page 48, line* − *15 (Inverse Image)*.
Herstein refers to the "inverse image of $\bar{g}$" where Halmos refers to the "inverse image of $\{\bar{g}\}$" (**Halmos**: *page 38*).

(viii) **Herstein**: *page 48, lines* − *13 to* − *5*.
These results are pictured below.

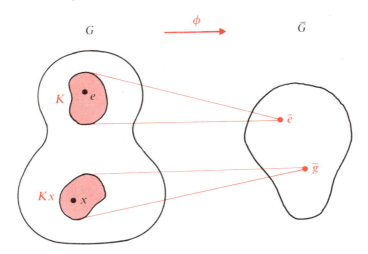

SAQ 6

Herstein: *page 56, Problem 12*.

(Solution is given on p. 39.)

SAQ 7

Let ϕ be a homomorphism with kernel K, from the group G onto the group $\bar{G}$.
Let $\bar{H}$ be a subgroup of $\bar{G}$ and let A be a subgroup of G.
Show that:

(i) $\phi^{-1}(\bar{H})$ is a subgroup of G;

(ii) K is a subgroup of $\phi^{-1}(\bar{H})$;

(iii) $\phi(A)$ is a subgroup of $\bar{G}$;

(iv) $\phi^{-1}(\phi(A)) = KA$.
    (HINT: Use Lemma 2.16.)

(Solution is given on p. 39.)

5.3   ISOMORPHISMS

5.3.0   Introduction

Isomorphisms are important examples of homomorphisms. Whereas homomorphisms may be many-to-one, isomorphisms must be one-to-one. Two groups are *isomorphic* if there is an isomorphism from one group *onto* the other. In this situation, the two group structures are mathematically indistinguishable. They can differ only in the way they are presented—in the set on which the group operation is specified, and the actual group operation.

For example, we know that the abstract group S_3 is isomorphic to the group of symmetry operations of an equilateral triangle. These represent two different ways of presenting the same group.

We are interested in the mathematical properties, and so we shall treat isomorphic groups as equivalent (once we have verified that "is isomorphic to" is an equivalence relation).

READ **Herstein**: *page 48, line* -1 *to page 49, line* -2.

Notes

(i)   *Herstein*: *page 49, line 7.*
It is worth emphasizing that an *isomorphism* is a one-to-one (homo)morphism. Two groups A, B are *isomorphic* if there is an isomorphism of A *onto* B. This is equivalent to the existence of a homomorphism $f: A \longrightarrow B$, and a homomorphism $g: B \longrightarrow A$, such that

$g \circ f: A \longrightarrow A$ is the identity mapping

and

$f \circ g: B \longrightarrow B$ is the identity mapping.

(In this functional notation, $f \circ g$ means first apply g, then f.)

(ii)   *Herstein*: *page 49, lines 10–12 (facts (1) to (3)).*
These are just the properties of an equivalence relation on the set of groups. To be precise, the collection of all groups is not a set, for it can be shown that any set can have a group structure defined on it, and so the set of all groups includes the notion of the set of all sets which we have ruled out. This difficulty can be overcome by considering only the subsets of a given set E, and the set of groups defined on subsets of E. To verify the three properties we observe:

(1) $\phi: G \longrightarrow G$ defined by $\phi(g) = g$, is an isomorphism of G onto G;

(2)   if $\phi: G \longrightarrow G^*$ is an isomorphism onto, then we can construct $\psi: G^* \longrightarrow G$ by $\psi(g^*) =$ the unique member of $\phi^{-1}(g^*)$. ψ is an isomorphism onto;

(3)   if $\phi: G \longrightarrow G^*$ is an isomorphism onto and $\psi: G^* \longrightarrow G^{**}$ is an isomorphism onto, then we have the following commutative diagram.

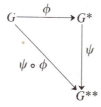

Clearly $\psi \circ \phi$ is an isomorphism onto. (Both one-to-one and onto are easy, but the fact that $\psi \circ \phi$ is a homomorphism should be checked.)

The fact that we have an equivalence relation on a suitable set of groups permits us to identify all those groups which are isomorphic, and treat them as being indistinguishable. What we do is to identify the equivalence classes with particular members of the equivalence class. Thus C_4, the cyclic group with 4 elements, represents the equivalence class of all groups isomorphic to C_4, including the integers under addition modulo 4. More generally, we can now refer to C_n as *the* cyclic group of order n.

(iii) **Herstein**: *page 49, lines* -7 *to* -2 (*Corollary and discussion*).
Note that in this corollary, ϕ is an isomorphism from G *into* $\bar{G}$, so that it does *not* necessarily follow that G is isomorphic to $\bar{G}$.

The technique of lines -5 to -2 is important. It will be used frequently in a slightly more general form to be discussed in the next section.

Examples

For each of the *Examples 0 to 5* in **Herstein**: *pages 46 to 47*, we shall determine which homomorphisms are isomorphisms. To do this we need only find the kernel in each case.

If $\phi(x) = e$ for all $x \in G$, the kernel $K_\phi = G$. Thus, we deduce the following facts.

0   ϕ is not an isomorphism unless G is the one-element group.

   If $\phi(x) = x$ for all $x \in G$, then the kernel $K_\phi = \{e\}$, so ϕ is an isomorphism.

1   ϕ is into. The kernel of ϕ is $\{0\}$, so ϕ is an isomorphism.

2   The kernel is $\{e, \psi, \psi^2\}$, so f is not an isomorphism.

3   ϕ is into. The kernel of ϕ is $\{0\}$, so ϕ is an isomorphism.

4   The kernel of ϕ is R^+, so ϕ is not an isomorphism.

5   The kernel of ϕ is $\{nt : t \in J\}$, so ϕ is not an isomorphism.

SAQ 8

Herstein: *page 55, Problem 2.*

(Solution is given on p. 40.)

SAQ 9

Herstein : *page 55, Problem 3.*
(Note: G is finite!)

(Solution is given on p. 40.)

5.3.1   The First Isomorphism Theorem

We know, by **Herstein**: *page 47, Lemma 2.13*, that there is a natural homomorphism from G onto G/N. The First Isomorphism Theorem demonstrates that any homomorphic image of G is isomorphic to a quotient group of G. Virtually the whole of Section 5.3.0 has been a preparation for this theorem.

This important theorem was proved in *Unit M100 33, Groups II*. All the constituent ideas have been treated in the earlier part of this unit, and it is now just a matter of accumulating the details.

First Isomorphism Theorem

Let ϕ be a homomorphism of G onto $\overline{G}$ with kernel K. Then

$$G/K \approx \overline{G}.$$

We now give a proof of this theorem.

If ϕ is a homomorphism from G *onto* $\overline{G}$ with kernel K, then we know that each member of $\overline{G}$ has as its inverse image a coset of K (*Lemma 2.16*, **Herstein**: *page 48*). Thus, there is a one-to-one correspondence between the cosets of K in G and the members of $\overline{G}$. But we have also seen that K is a normal subgroup of G (*Lemma 2.15*, **Herstein**: *page 48*), and so we can form the quotient group G/K, where the elements of *this* group are the cosets of K. Thus we have established a one-to-one correspondence between the groups G/K and $\overline{G}$.

Herstein denotes this one-to-one and onto function by $\psi : G/K \longrightarrow \overline{G}$, where ψ maps the coset Kg to $\phi(g)$. That is, each coset is made to correspond to the obvious member of $\overline{G}$, namely the single element $\phi(g)$ which is the image of all members of Kg under the action of ϕ.

We know two other things: the group operation in G/K is given by

$$KgKh = Kgh;$$

and, since ϕ is a homomorphism,

$$\phi(g)\phi(h) = \phi(gh).$$

Therefore, it is clear that the elements combine in $\overline{G}$ by imitating the group operation in G/K.

The function ψ is, of course, a homomorphism:

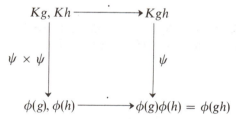

We have already pointed out that ψ is one-to-one and onto, so ψ is in fact an isomorphism. Thus the groups G/K and $\overline{G}$ are isomorphic.

In the proof of Theorem 2.D, Herstein chooses to put the pieces together in a slightly different order and he also proves several things that he could have got more simply by appealing to previous results. First, he defines ψ and proves that he has indeed defined a function. He then proves that ψ is onto, that it is a homomorphism and, lastly, that it is one-to-one. One thing that Herstein does gain from his proof of the theorem is the opportunity to use the corollary of *page 49*, demonstrating, in a fairly simple situation, the technique of proving that a particular homomorphism is in fact an isomorphism by showing that the kernel is just the identity element in G.

READ **Herstein**: *page 50 (the top) to page 51, line* -7.

The importance of the First Isomorphism Theorem cannot be over-emphasized. Herstein summarizes this in two places:

Herstein: *page 49, last paragraph and* **Herstein**: *page 51, lines 20–30 (middle paragraph)*.

Example

Let $(R, +)$ denote the additive group of the reals, $(J, +)$ denote the additive group of the integers. Let $(C, \cdot)$ denote the "multiplicative group of the circle" in which

$$C = \{e^{ix} : x \in R\}$$

and $e^{ix} = 1$ if and only if $x = 2n\pi$, n an integer. (See *Unit M100 29, Complex Numbers II.*)

Show that $(R, +)/(J, +) \approx (C, \cdot)$.

Solution

Before exhibiting the desired isomorphism, let us investigate $(R, +)/(J, +)$ briefly. The elements (cosets of $(J, +)$ in $(R, +)$) of this quotient group are the sets

$$a + J = \{(a + n) \in R : a \in R, n \in J\} \qquad (a \text{ fixed, } n \text{ variable}).$$

Thus, by considering all a in the interval $[0, 1)$, we obtain all the distinct elements in the quotient group. That is, each element of the quotient group has a representative in the interval $[0, 1)$. Furthermore, the end-points 0 and 1 are identified because they are in the same coset. This suggests that the effect of forming the quotient group can be depicted geometrically by mapping $[0, 1)$ of R onto a circle, as shown in the following diagram.

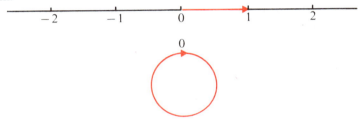

Similarly, any interval $[n, n + 1)$ of R is mapped onto the same circle.

Now, we note that the mapping

$$x \longmapsto e^{ix}$$

maps all intervals $[2\pi n, 2\pi(n + 1))$ of R onto a circle of radius 1 in the complex plane; so, if we account for a scaling factor of 2π, it seems reasonable that $(R, +)/(J, +)$ is isomorphic to $(C, \cdot)$. We shall now show this in detail.

First, we construct a homomorphism

$$\phi : (R, +) \longrightarrow (C, \cdot)$$

which is onto, and then we check to find its kernel. By the *Corollary*, **Herstein**: *page 49*, we want the kernel of ϕ to be J, i.e., the identity of $(R, +)/(J, +)$. So, since the identity in $(C, \cdot)$ is $e^{i0} = 1$, we want

$$\phi(n) = 1 \qquad \text{for all } n \in J.$$

A suitable ϕ is specified by

$$\phi : x \longmapsto e^{2i\pi x}.$$

(Note that 2π serves the role as the scale factor.) It is easy to show that ϕ is a homomorphism, since

$$\phi(x + y) = e^{2i\pi(x + y)} = e^{2i\pi x} e^{2i\pi y} = \phi(x)\phi(y).$$

ϕ is clearly onto, since for any y, if we want $\phi(x) = e^{iy}$, we just choose $x = y/2\pi$. Finally,

$$\text{the kernel of } \phi = \{x : \phi(x) = 1\}$$
$$= \{x : e^{2i\pi x} = 1\}$$
$$= \{x : 2\pi x = 2\pi n \text{ for some integer } n\}$$
$$= J.$$

Then the First Isomorphism Theorem tells us that

$$(R, +)/(J, +) \approx (C, \cdot).$$

5.3.2 Cauchy's Theorem for Finite Abelian Groups

Let G be a finite Abelian group of order n. Cauchy's Theorem for finite Abelian groups states that, for each prime number p which divides n, there is a cyclic subgroup of G of order p.

For example, C_{210}, the cyclic group of order 210, must have cyclic subgroups of orders 2, 3, 5 and 7.

The importance of this theorem is that it is also valid for finite non-Abelian groups (although the proof is harder, and will be delayed until *Unit 12*). At the moment, it is a good application of the group-theoretic ideas developed so far. The structure of the proof is typical of many proofs in finite groups. You should take care to observe the *strategy* of the proof, which is much more important than the details. Group theorists do not memorize proofs—they accumulate strategies.

Herstein's proof is rather involved.

We provide a flow chart of the argument. You may find it useful to examine this first. Otherwise read the proof and then look at the flow chart if you get stuck. The overall strategy of the proof is to use mathematical induction on $o(G)$.

*READ **Herstein**: page 51 (Application 1) to page 52 (end of Application 1).*

Notes

The strategy of the proof is demonstrated by the following flow chart.

(i) *page 51, line −3*
     to page 52, line 2

(ii) *page 52, line 3*

(iii) *page 52, lines 3 to 6*

(iv) *page 52, line 6*

(v) *page 52, line 6*

(vi) *page 52, lines 7 to 9*
      $p \nmid o(N)$ means p does not
      divide $o(N)$.

(vii) *page 52, line 10*
       G/N is Abelian by
       SAQ 6, page 14.

(viii) *page 52, lines 10 to 12*

(ix) *page 52, lines 12 to 13*

(x) *page 52, lines 14 to 17*
     For the appropriate corollary
     to Lagrange's Theorem, see
     ***Herstein**: page 37, Corollary 2.*

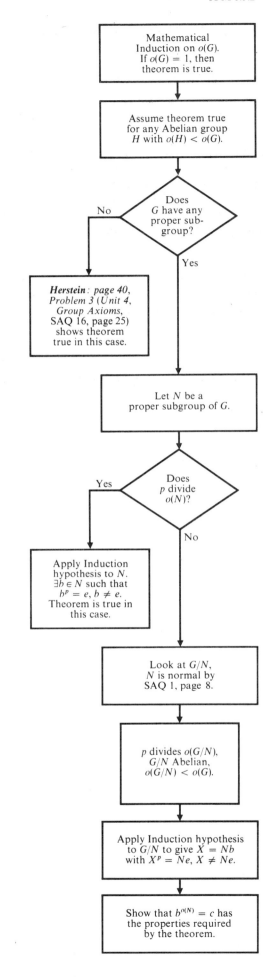

Mathematical
Induction on $o(G)$.
If $o(G) = 1$, then
theorem is true.

Assume theorem true
for any Abelian group
H with $o(H) < o(G)$.

Does
G have any
proper sub-
group?

No

Yes

***Herstein**: page 40,
Problem 3 (Unit 4,
Group Axioms,
SAQ 16, page 25)
shows theorem
true in this case.*

Let N be a
proper subgroup of G.

Does
p divide
$o(N)$?

Yes

No

Apply Induction
hypothesis to N.
$\exists b \in N$ such that
$b^p = e, b \neq e$.
Theorem is true in
this case.

Look at G/N,
N is normal by
SAQ 1, page 8.

p divides $o(G/N)$,
G/N Abelian,
$o(G/N) < o(G)$.

Apply Induction hypothesis
to G/N to give $X = Nb$
with $X^p = Ne, X \neq Ne$.

Show that $b^{o(N)} = c$ has
the properties required
by the theorem.

The last box is the crux of the proof. In detail, the corresponding argument is as follows.

To show that $c \neq e$, we argue by contradiction, and assume that $c = e$. Then

$$b^{o(N)} = e,$$

and so

$$(Nb)^{o(N)} = N.$$

But we also know that

$$(Nb)^p = N, \; p \nmid o(N).$$

We use the standard argument that, since $(p, o(N)) = 1$, there exist integers λ and μ such that

$$\lambda p + \mu o(N) = 1.$$

Then

$$(Nb)^{\lambda p + \mu o(N)} = Nb.$$

But

$$
\begin{aligned}
(Nb)^{\lambda p + \mu o(N)} &= ((Nb)^p)^\lambda ((Nb)^{o(N)})^\mu \\
&= N^\lambda N^\mu \\
&= N,
\end{aligned}
$$

so $Nb = N$ and $b \in N$. This contradicts the statement that $b \notin N$. Therefore, $c \neq e$ and the proof is complete.

Note that although the strategy of the proof is a reasonable one, the proof relies heavily on several non-trivial results such as:

1   G has no non-trivial subgroups $\Rightarrow$ G is of prime order.

2   G is Abelian $\Rightarrow$ every subgroup of G is normal.

3   G is Abelian $\Rightarrow$ G/N is Abelian.

4   N is a subgroup of G $\Rightarrow$ $o(N)|o(G)$ (Lagrange's Theorem).

5   $a \in G \Rightarrow a^{o(G)} = e$ (Corollary to Lagrange's Theorem).

Thus, our strategy is an effective one because we have armed ourselves with the appropriate strategies for carrying out the details of the proof.

SAQ 10

What feature of Abelian groups, not true in general for non-Abelian groups, was used in the proof?

(Solution is given on p. 41.)

5.3.3 Sylow's Theorem for Finite Abelian Groups

Let G be a finite Abelian group of order n. Sylow's Theorem for Abelian groups assures us that, if p^α is the largest power of the prime p which divides $o(G) = n$, then there is a subgroup of G of order p^α. Such a subgroup is called a *p-Sylow subgroup* of G.

We know from Lagrange's Theorem that the only possible orders for subgroups are numbers dividing $o(G)$. So the theorem actually *guarantees* the existence of certain special subgroups of G. That is, we have a *partial converse* of Lagrange's Theorem.

We do not ask you to read the proof of Sylow's Theorem in the present form.

READ Herstein: page 52, the statement of Application 2, and page 53, lines 7 to 20.

A more general form of Sylow's Theorem will be given in *Unit 12* of this course. The more general Sylow Theorems, which apply to all groups, permit the analysis of the structure of the subgroups of many groups whose orders only are known.

Examples

Let G be the additive group J_n of integers modulo n.

If $n = 14$, there exist subgroups of order 2 and 7. Call these G_2 and G_7 respectively (the 2-Sylow and 7-Sylow subgroups). Then $G_2 \cap G_7 = \{e\}$, since any element in $G_2 \cap G_7$ must have order dividing both 2 and 7.

Now G is Abelian, so by SAQ 1, page 8, G_2 and G_7 are normal subgroups of G. Therefore, by SAQ 5, page 11, G_2G_7 is a subgroup of G, and by *Herstein: page 39, Theorem 2.B*, G_2G_7 has order

$$\frac{o(G_2)o(G_7)}{o(G_2 \cap G_7)} = 14 = o(J_{14}) = o(G).$$

Thus

$$G = G_2G_7.$$

We have a fair amount of information about J_{14}, using only the information that it is Abelian, and has order 14.

If $n = 36$, there exist subgroups of orders 4 and 9. Call these G_4 and G_9. Again,

$$G_4 \cap G_9 = \{e\}$$

and

$$J_{36} = G = G_4G_9.$$

We have not determined G_4 or G_9. Without using any further information, we cannot determine the structure of G_4 or G_9. We do know, however, that J_{36} is cyclic. Therefore G_4 and G_9 must also be cyclic. (See *Unit 4, Group Axioms*, SAQ 17, page 25.)

5.3.4 The Second Isomorphism Theorem

This theorem tells us how to take quotient groups of quotient groups. It does so by establishing a one-to-one mapping between the subgroups of G which contain the normal subgroup N, and the subgroups of G/N. You will get more benefit from this section if you follow the development below, working as you go, rather than reading straight through the section. The part of **Herstein** covered is *page 53, line 21 to the end of the section.*

LEMMA 2.17

Let ϕ be a homomorphism of G onto $\bar{G}$ with kernel K, and let $\bar{H}$ be a subgroup of $\bar{G}$.

$$H = \phi^{-1}(\bar{H}) = \{x \in G : \phi(x) \in \bar{H}\}.$$

Then

(i) H is a subgroup of G with $K \subset H$;

(ii) $\bar{H}$ normal in $\bar{G} \Leftrightarrow H$ normal in G;

(iii)   there is a one-to-one correspondence between subgroups of G containing K and subgroups of $\bar{G}$.

PROOF

(i)   This was proved in SAQ 7, page 14.

(ii)   If $\bar{H}$ is normal in $\bar{G}$, then we want to prove that $\forall g \in G, gHg^{-1} \subset H$.

Now $\phi(gHg^{-1}) = \phi(g)\bar{H}\phi(g^{-1}) = \phi(g)\bar{H}\phi(g)^{-1} = \bar{H}$ since $\bar{H}$ is normal in $\bar{G}$.

Therefore $gHg^{-1} \subset H$ and H is normal.

If H is normal in G, then

$$\phi(g)\phi(H)\phi(g^{-1}) = \phi(gHg^{-1}) = \phi(H) = \bar{H}.$$

And, since ϕ is onto, every element in $\bar{G}$ can be written as $\phi(g)$ for some $g \in G$. Thus $\bar{H}$ is normal in $\bar{G}$.

(iii)   The one-to-one correspondence is implicit in ϕ. With each subgroup H of G containing K, we associate $\phi(H)$. This association is one-to-one because if

$$\phi(H_1) = \phi(H_2)$$

then

$$\phi^{-1}(\phi(H_1)) = \phi^{-1}(\phi(H_2)),$$

and so by SAQ 7, page 14,

$$KH_1 = KH_2.$$

Since, by hypothesis, $K \subset H_1$ and $K \subset H_2$, it follows that

$$H_1 = H_2.$$

$$
\begin{array}{ccc}
G & \xrightarrow{\phi} & \bar{G} \\
\vert & & \vert \\
H & \longrightarrow & \bar{H} \\
\vert & & \vert \\
K & \longrightarrow & \langle \bar{e} \rangle
\end{array}
$$

THEOREM 2.E (The Second Isomorphism Theorem).

Let ϕ be a homomorphism of G onto $\bar{G}$ with kernel K.

Let $\bar{N}$ be a normal subgroup of $\bar{G}$. Put

$$N = \phi^{-1}(\bar{N}).$$

Then

$$G/N \approx \bar{G}/\bar{N},$$

that is,

$$G/N \approx (G/K)/(N/K).$$

PROOF

We shall now show that the mapping

$$\sigma : G/N \longrightarrow \bar{G}/\bar{N}$$

specified by

$$Ng \longmapsto \bar{N}\phi(g),$$

is, as may be expected, an isomorphism. We must check that σ is:

(i)   well-defined (independent of the particular coset representative used)

(ii)   a homomorphism

(iii) onto

(iv) a morphism whose kernel is N, the identity of G/N.

Each of these is straightforward:

(i)   If $Ng_1 = Ng_2$, then $g_1 = ng_2$ for some $n \in N$. Then

$$\sigma(Ng_1) = \overline{N}\phi(g_1)$$
$$\sigma(Ng_2) = \overline{N}\phi(g_2).$$

But

$$\sigma(Ng_1) = \overline{N}\phi(g_1) = \overline{N}\phi(ng_2) = \overline{N}\phi(g_2) = \sigma(Ng_2).$$

Thus σ is indeed independent of the particular coset representative and so is well-defined.

(ii) $\sigma(Ng_1 Ng_2) = \sigma(Ng_1 g_2)$

$$= \overline{N}\phi(g_1 g_2)$$
$$= \overline{N}\phi(g_1)\phi(g_2)$$
$$= \overline{N}\phi(g_1)\overline{N}\phi(g_2)$$
$$= \sigma(Ng_1)\sigma(Ng_2).$$

(iii) For any $\overline{N}y \in \overline{G}/\overline{N}$, $y = \phi(x)$ for some $x \in G$. Thus

$$\sigma(Nx) = \overline{N}\phi(x) = \overline{N}y,$$

and σ is onto.

(iv) Let the kernel of σ be denoted by K_σ. To show that $K_\sigma = N$, note that if Ng is any element in K_σ, then

$$\sigma(Ng) = \overline{N}\phi(g) = \overline{N}$$

and so $\phi(g) \in \overline{N}$. But, by the definition of N, this means that $g \in N$. Consequently,

$$Ng = N$$

and N is the only element in K_σ, completing the proof.

Thus,

$$G/N \approx \overline{G}/\overline{N}.$$

We can immediately apply our theorem to obtain an equivalent result expressed in terms of quotient groups.

If a group G has a normal subgroup K, then we have a homomorphism $\phi : G \longrightarrow G/K$, where $\phi(g) = Kg$. Now let $\overline{N}$ be a normal subgroup of G/K. Then $N = \phi^{-1}(\overline{N})$ is normal in G, and of course $\overline{N} = N/K$. (See SAQ 5(ii).) We now have exactly the conditions of the theorem with G/K and N/K replacing $\overline{G}$ and $\overline{N}$ respectively. It follows that

$$G/N \approx (G/K)/(N/K).$$

Herstein's approach is to use the First Isomorphism Theorem to establish his isomorphisms, rather than construct them directly. His homomorphisms are given by the diagram

He then shows that the kernel of ψ is N and so

$$G/N \approx \overline{G}/\overline{N}.$$

SAQ 11
Herstein*: page 55, Problem 7.*
The group operation is composition.

(Solution is given on p. 41.)

5.3.5 The Third Isomorphism Theorem

This theorem relates two different quotient groups by an isomorphism. It would be most valuable if you can try to prove the theorem yourself. It is stated in ***Herstein****: page 55, Problem 6*. If you get stuck, come back to the commentary for hints. There are no new ideas involved. The strategy we recommend is to construct a homomorphism from N onto NM/M and then use the First Isomorphism Theorem. You can try to construct the isomorphism directly, but showing that it is well-defined is a bit tricky.

THIRD ISOMORPHISM THEOREM
If N and M are normal subgroups of G, then

$$NM/M \approx N/(N \cap M).$$

PROOF
The strategy is to specify a homomorphism from one of the numerator groups onto the other quotient group. We then check that the kernel is as required. There are two choices possible:

$$\phi : N \longrightarrow NM/M \text{ and } \psi : NM \longrightarrow N/(N \cap M).$$

The second involves complicated calculation. The first is done as follows.

The group $NM/M = \{Mx : x \in NM\}$. But since $NM = MN$ (see SAQ 5, page 11),

$$x \in NM \Rightarrow x = mn \text{ for some } m \in M, n \in N$$

$$\Rightarrow Mx = Mmn = Mn.$$

Thus, we can clearly choose our coset representative to be in N.

A reasonable specification for ϕ is

$$\phi : N \longrightarrow MN/M$$

where
$$\phi : x \longmapsto Mx.$$

Is ϕ a homomorphism? Yes, since

$$\phi(xy) = Mxy$$

$$= (Mx)(My) \qquad (M \text{ is normal in } MN)$$

$$= \phi(x)\phi(y).$$

Is ϕ onto? Yes, since $MN/M = \{Mx : x \in N\}$.

What is the kernel of ϕ?

$$K_\phi = \{x : Mx = M, \ x \in N\}$$

$$= \{x : x \in M \text{ and } x \in N\}$$

$$= N \cap M.$$

Thus

$$N/(N \cap M) \approx NM/M,$$

by the First Isomorphism Theorem.

5.3.6 Summary

To establish an isomorphism between a quotient group G/N and a group H the best way is usually to find a homomorphism from G onto H whose kernel is N. The First Isomorphism Theorem then guarantees that

$$G/N \approx H.$$

Care must be taken, when specifying the homomorphism, to ensure that it is well-defined.

If G is a finite *Abelian* group, with p^α dividing $o(G)$ but $p^{\alpha+1}$ not dividing $o(G)$, then Cauchy's Theorem and Corollary says that there are elements of G with order p (cyclic subgroups of order p) and Sylow's Theorem says that there is a subgroup of order p^α as well.

The Second and Third Isomorphism Theorems are less important than the first. The second states that

$$G/N \approx (G/K)/(N/K),$$

where K is a normal subgroup of G, contained in the normal subgroup N. The third states that

$$NM/M \approx N/(N \cap M),$$

where N and M are normal in G.

5.4   GENERATED SUBGROUPS AND THE COMMUTATOR SUBGROUP

5.4.0   Introduction

This section is based entirely on the contents of two problems from ***Herstein***—*Problems 4 and 5 on page 55*. The first of these describes how a set of elements in a group can be said to generate a subgroup. This idea is extremely far-reaching, and is basic to the topological applications of group theory. The second problem uses this construction to discuss the commutator group.

The idea of a generated subgroup is very straightforward. If U is a subset of a group G, we want to extend U to a subgroup of G, so we begin by forming the product in G of any two elements of U, and adding these to U. We must also put the inverse of each element of U into U. This process makes U larger. It may not be a group yet, but we can continue this process until we do get a subgroup. The result, which we write as

$$\langle U \rangle$$

is "generated" by all possible products of elements of U, and their inverses. (Our notation is different from that of Herstein, who denotes this subgroup by $\hat{U}$. We have used $\langle U \rangle$ to conform to the majority of texts on group theory.) There is a close connection with this idea and that of the span of a set of vectors in a vector space.

Starting with an arbitrary group, it is frequently useful to examine not the whole group, but an Abelian homomorphic image. After all, Abelian groups are much easier to deal with than non-Abelian groups. The *commutator subgroup* of a group is the smallest normal subgroup such that the quotient group is Abelian.

The name *commutator* is derived from the form of the generating elements of the subgroup. If G is Abelian, then

$$xy = yx,$$

which can be written as

$$xyx^{-1} = y \quad \text{or} \quad xyx^{-1}y^{-1} = e.$$

On the other hand, if G is not Abelian, $xyx^{-1}y^{-1}$ may not be equal to e, but can be used to measure "how far" G is from being Abelian. Any element of G which can be written as $xyx^{-1}y^{-1}$ is called a *commutator*. By factoring out a normal subgroup which contains all the commutators, we remove all of the non-Abelian-ness of G (if any), and so we expect the quotient group to be Abelian.

You should now be in a position to tackle ***Herstein***: *Problems 4 and 5, page 55*, on your own. If you are short of time, you can simply read the next two sections, but you will not gain as much as if you try the problems first. In any case, you should read the next two sections at some time, because they contain additional properties not mentioned in ***Herstein***.

5.4.1   Generated Subgroups

*Example 1 (**Herstein**: page 55, Problem 4a).*

Given any group G and a subset U, let $\langle U \rangle$ be the smallest subgroup of G which contains U. Prove that there is such a subgroup $\langle U \rangle$ of G. $\langle U \rangle$ is called the *subgroup generated by* U.

Solution 1

The easiest approach is to look at

$$S = \{H : U \subset H \text{ and } H \text{ is a subgroup of } G\}.$$

Now $S \neq \varnothing$, since $G \in S$. We are looking for a member of S which is smaller than (contained in) all other members. We look at the intersection of all the members of S. Let

$$\langle U \rangle = \bigcap S = \bigcap_{H \in S} H.$$

Now $\langle U \rangle$ is a subgroup of G, because if $x, y \in \langle U \rangle$, then for any $H \in S$,

$$x, y \text{ and } xy \text{ are all members of } H.$$

Thus $\langle U \rangle$ is closed under group multiplication. Similarly, for each $x \in \langle U \rangle$, if $H \in S$,

$$x \text{ and } x^{-1} \text{ are members of } H$$

and so $x^{-1} \in \langle U \rangle$. Thus $\langle U \rangle$ is a subgroup of G.

We also know that $U \subset \langle U \rangle$ since $U \subset H$ for each H in S. Thus $\langle U \rangle \in S$, and clearly any subgroup of G which contains U also contains $\langle U \rangle$. Therefore $\langle U \rangle$ has the properties claimed.

An alternative approach which indicates the source of the words "generated by," is to form a recursive definition of $\langle U \rangle$ as follows.

Let

$$U_0 = U \subset G.$$

Let

$$U_{n+1} = U_n \cup \{xy : x, y \in U_n\} \cup \{x^{-1} : x \in U_n\}.$$

(Here we throw in all the products and inverses of elements in U_n.)

Let

$$\langle U \rangle = \bigcup_{n \in \omega} U_n.$$

It is easy to check that $\langle U \rangle$ is a subgroup of G, and of course contains U. It is not so easy to see that $\langle U \rangle$ is the smallest such subgroup, until you realize that if H is any subgroup of G containing U, then

$$U_0 \subset H$$

and, if $U_n \subset H$ then $U_{n+1} \subset H$ since H is a group.

Thus, by mathematical induction, $\langle U \rangle \subset H$.

Note that, if $x \in \langle U \rangle$, then $x = a_1 a_2 a_3 \ldots a_n$ for some $n \in \omega$, with a_i or $a_i^{-1} \in U$ for each i. The reason for this is that x is a product of a product of ... a product of elements of U and their inverses.

The difference in the approaches is that the first isolates the subset $\langle U \rangle$ common to all subgroups of G containing U, and the second builds U up into a subgroup of G, by putting in only the necessary elements to make the result a subgroup.

Example 2 (**Herstein**: *page 55, Problem 4b and its extension*).

(i)   If $gug^{-1} \in U$ for all $g \in G$, $u \in U$, show that $\langle U \rangle$ is a normal subgroup of G.

(ii) Show that there exists a subgroup $\langle U \rangle^N$, called the *normal closure* of U, normal in G, which contains U, and is the smallest such subgroup of G.

Solution 2

(i)   If $gug^{-1} \in U$ for all $g \in G$, $u \in U$, consider a general element $x \in \langle U \rangle$. Now x must be a product of the form

$$x = a_1 \ldots a_n,$$

where for each i, $1 \leqslant i \leqslant n$,

$$a_i \text{ or } a_i^{-1} \in U.$$

Now

$$gxg^{-1} = g(a_1 \ldots a_n)g^{-1}$$
$$= (ga_1g^{-1})(ga_2g^{-1}) \ldots (ga_ng^{-1}).$$

Since a_i or $a_i^{-1} \in U$, it follows from the hypothesis that either

$$ga_ig^{-1} \quad \text{or} \quad (ga_ig^{-1})^{-1} \in U,$$

and so $gxg^{-1} \in \langle U \rangle$. Thus under the hypothesis that $gug^{-1} \in U$ for each $u \in U$, $\langle U \rangle$ is a normal subgroup of G.

(ii) We want to extend U to have the property of part (i). To this end, we put

$$V = \{gug^{-1} : g \in G, u \in U\}.$$

Notice that $U \subset V$ since $eue^{-1} = u$, $\forall u \in U$. Now, for any $v \in V$, there exists $g \in G, u \in U$ such that $v = gug^{-1}$, so that, for any $x \in G$,

$$xvx^{-1} = xgug^{-1}x^{-1} = (xg)u(xg)^{-1},$$

and so

$$xvx^{-1} \in V.$$

Therefore, V has the property of part (i). Thus $\langle V \rangle$ is normal in G because of part (i), and $U \subset V$. The only problem now is to show that V is the smallest such subgroup of G. Now any normal subgroup M of G which contains U must contain V, by the definition of normality. Therefore M must contain $\langle V \rangle$, and we have shown that $\langle U \rangle^N = \langle V \rangle$.

Alternatively, we could have observed that the intersection of all normal subgroups of G which contain U forms a normal subgroup of G. Reasoning as in Example 1, this intersection would just be the required normal closure $\langle U \rangle^N$.

SAQ 12

In the group A_4, find the normal closure of the singleton $\{c\}$.

(Solution is given on p. 42.)

5.4.2   The Commutator Subgroup

In this section, we shall investigate several properties of the commutator subgroup of a group. Our investigation will involve most of the ideas discussed in the unit so far. By attempting to prove the assertions below and by following our solutions, you should obtain valuable experience in manipulating the group-theoretic concepts involved. However, you are not meant to memorize the results.

*Example (**Herstein**: page 55, Problem 5 and extensions).*

Let $U = \{xyx^{-1}y^{-1} : x, y \in G\}$. Let $G' = \langle U \rangle$ (the *commutator* subgroup of G).

(i)   Find S_3'. (HINT: use the cards. There is no need to work out all 36 commutators —try to cut down on the number of calculations you need.)

(ii)   (a)   Show that G' is normal in G.
     (b)   Show that G/G' is Abelian.
     (c)   If G/N is Abelian, show that $G' \subset N$.
     (d)   Show that if H is a subgroup of G with $G' \subset H$, then H is a normal subgroup of G.
     (e)   Let $\mathscr{G}$ be the collection of groups defined on the subsets of some set E. We know that to each $G \in \mathscr{G}$, there corresponds a group G'. In other words:

$$' : \mathscr{G} \longrightarrow \mathscr{G}$$

$$' : G \longmapsto G'.$$

Let ϕ be a homomorphism from G onto a group H. Show that $\phi(G') = (\phi(G))'$, that is, that the following is a commutative diagram:

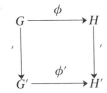

where ϕ' is just ϕ restricted to G'.

     (f)   Show that if $\phi : G \longrightarrow H$ is onto, then there is a natural way to specify a homomorphism

$$\bar{\phi} : G/G' \longrightarrow H/H', \qquad \text{where } H' = \phi(G'),$$

so that for this $\bar{\phi}$,

$$G \xrightarrow{\phi} H$$

is a commutative diagram.

(iii) Using (a) and (c), find A_4' without direct calculation.

Solution

(i) S_3 has 6 elements. There are therefore $6 \times 6 = 36$ commutators $(ghg^{-1}h^{-1})$. Notice that

$$(ghg^{-1}h^{-1})^{-1} = hgh^{-1}g^{-1},$$

so we need only calculate one of these.

Also if g or $h = e$, then $ghg^{-1}h^{-1} = e$. Therefore we need only calculate $\dfrac{5 \times 4}{2} = 10$ products.

These are

$$\psi\phi\psi^{-1}\phi^{-1} = \psi^2 = \psi^{-1} = \psi^2$$

$$\psi(\psi\phi)\psi^{-1}(\psi\phi)^{-1} = \psi^2$$

$$\psi(\psi^2\phi)\psi^{-1}(\psi^2\phi)^{-1} = \psi^2$$

$$\psi(\psi^2)\psi^{-1}(\psi^2)^{-1} = e$$

$$\psi^2\phi(\psi^2)^{-1}\phi^{-1} = \psi$$

$$\psi^2(\psi\phi)(\psi^2)^{-1}(\psi\phi)^{-1} = \psi$$

$$\psi^2(\psi^2\phi)(\psi^2)^{-1}(\psi^2\phi)^{-1} = \psi$$

$$(\psi\phi)(\psi^2\phi)(\psi\phi)^{-1}(\psi^2\phi)^{-1} = \psi$$

$$(\psi\phi)(\phi)(\psi\phi)^{-1}\phi^{-1} = \psi^2$$

$$(\psi^2\phi)\phi(\psi^2\phi)^{-1}\phi^{-1} = \psi.$$

Thus, the commutators are just $\{e, \psi, \psi^2\}$. This set is already a subgroup of S_3 and so it is the commutator subgroup. Note: for very few groups is it the case that the commutators alone form a subgroup.

The properties in part (ii) will permit us to calculate the commutator subgroup very easily, without all this calculation.

(ii) (a) Applying the result of **Herstein**: *page 55, Problem 4(b)*, we observe that for any

$$u = xyx^{-1}y^{-1} \in U \text{ and } g \in G,$$

$$g(xyx^{-1}y^{-1})g^{-1} = (gxg^{-1})(gyg^{-1})(gx^{-1}g^{-1})(gy^{-1}g^{-1})$$

$$= (gxg^{-1})(gyg^{-1})(gxg^{-1})^{-1}(gyg^{-1})^{-1}$$

$$\in G'.$$

Therefore G' is normal in G.

(b) You could do messy computations with cosets, but there is no need ever to do this. We work entirely with morphisms.

As in **Herstein**: *page 47, Lemma 2.13*, let

$$\phi : G \longrightarrow G/G'.$$

Then ϕ is onto and so every element of G/G' is of the form $\phi(x)$. Examining a commutator in G/G', we find

$$\phi(x)\phi(y)\phi(x)^{-1}\phi(y)^{-1} = \phi(x)\phi(y)\phi(x^{-1})\phi(y^{-1})$$

$$= \phi(xyx^{-1}y^{-1})$$

$$= \bar{e}, \text{ the identity in } G/G'.$$

Thus all commutators in G/G' are trivial, which means that G/G' is Abelian.

(c) Let $\phi : G \longrightarrow G/N$ be onto. For any $x, y \in G$,

$$\phi(xyx^{-1}y^{-1}) = \phi(x)\phi(y)\phi(x)^{-1}\phi(y)^{-1}$$

$$= \phi(x)(\phi(x))^{-1}\phi(y)(\phi(y))^{-1} \text{ since } G/N \text{ is Abelian}$$

$$= \bar{e}.$$

Thus $G' \subset N$.

(d) If $G' \subset H$, a subgroup of G, then by SAQ 5(ii) page 11 H/G' is a subgroup of G/G', which is Abelian by part (b) above. Therefore, since every subgroup of an Abelian group is normal,

$$H/G' \text{ is normal in } G/G'.$$

By Lemma 2.17, H is normal in G.

This can also be done by direct calculation.

For any $x \in G, h \in H, xhx^{-1} = (xhx^{-1}h^{-1}h)$. This element is in H because $xhx^{-1}h^{-1} \in G' \subset H$.

Thus H is normal in G.

(e) Since

$$\phi(G) = H = \{\phi(x) : x \in G\}$$

and

$$(\phi(G))' = H' = \langle \{pqp^{-1}q^{-1} : p, q \in H\} \rangle$$

we have that

$$H' = \langle \{\phi(x)\phi(y)(\phi(x))^{-1}(\phi(y))^{-1} : x, y \in G\} \rangle.$$

Thus, H' is the set generated by products and inverses of elements of the form

$$\phi(x)\phi(y)(\phi(x))^{-1}(\phi(y))^{-1}.$$

Now,

$$\phi(G') = \{\phi(s) : s \in G'\}$$

and each such $s \in G'$ is a product of elements of the form

$$xyx^{-1}y^{-1}, x, y \in G$$

and their inverses. So because

$$\phi(xyx^{-1}y^{-1}) = \phi(x)\phi(y)(\phi(x))^{-1}(\phi(y))^{-1},$$

$\phi(G')$ is also the set generated by products and inverses of elements of the form

$$\phi(x)\phi(y)(\phi(x))^{-1}(\phi(y))^{-1}.$$

Thus
$\phi(G') = (\phi(G))'$ and the given diagram is commutative. Note that, in general, G' is properly contained in G, so we must define ϕ' as the restriction of ϕ to G'.

(f) Elements of G/G' are of the form $G'x$.

Now $\phi(x) \longmapsto H'\phi(x)$ under the homomorphism $G \longrightarrow G/G'$, so we want $\bar{\phi}(G'x) = H'\phi(x)$.

$$\begin{array}{ccc}
x & \xrightarrow{\phi} & \phi(x) \\
\downarrow & & \downarrow \\
G'x & \xrightarrow{\bar{\phi}} & H(\phi(x))
\end{array}$$

However, we must show that $\bar{\phi}$ is well-defined. We have specified $\bar{\phi}$ in terms of a coset representative. If we change representatives, we must check that $\bar{\phi}$ still gives the same answer.

Now if $G'x = G'y$ we want $H'\phi(x) = H'\phi(y)$.

But $G'x = G'y$ means that $x = gy$ with $g \in G'$.

Then $H'\phi(x) = H'\phi(gy) = H'\phi(g)\phi(y) = H'\phi(y)$ since $g \in G'$ implies $\phi(g) \in H$.

Therefore $\bar{\phi}$ is a well-defined function. The same reasoning yields $\bar{\phi}$ as a homomorphism, since

$$\bar{\phi}(G'x \, G'y) = \bar{\phi}(G'xy)$$
$$= H'\phi(xy)$$
$$= H'\phi(x)\phi(y)$$
$$= H'\phi(x) \, H'\phi(y)$$
$$= \bar{\phi}(G'x) \, \bar{\phi}(G'y).$$

Now $\bar\phi$ was constructed so that the diagram is commutative, so our proof is complete.

(iii) We know that A_4 has only one normal subgroup, $N = \{e, a, b, ab\}$. Since A_4 is not Abelian, $A_4' \neq \{e\}$. And since A_4' is normal (by (a)), A_4' must therefore be N or A_4.

We must decide which. Now A_4/N has order 3 and is thus Abelian. So, by (c), $A_4' \subset N$, whence $A_4' = N$.

5.5   SUMMARY OF THE TEXT

In this unit you have been reminded of the importance of morphisms when analysing algebraic structure. The construction of quotient groups is equivalent (via the First Isomorphism Theorem) to the finding of homomorphic images. The notion of a normal subgroup is closely related to the kernel of a homomorphism.

These ideas were used to prove two important theorems about finite Abelian groups (Cauchy and Sylow) which guarantee the existence of certain subgroups of a finite group.

The Second and Third Isomorphism Theorems then demonstrated the properties of quotients of quotients, and the interrelation between two normal subgroups.

The really important ideas are the definition of a normal subgroup, and the First Isomorphism Theorem. From the definition of a normal subgroup, the other basic properties can be easily checked, once you are familiar with how easily expressions of the form gxg^{-1} can be manipulated. The First Isomorphism Theorem underlines the usefulness of using homomorphisms to get at properties of groups.

The last section dealt with the concept of a subset of a group generating a subgroup, and the commutator subgroup as a particular case (and by far the most important one). You will see these ideas carried further in a later unit.

The following terms have been introduced in this text:

Commutator Subgroup
Generated Subgroup
Homomorphism
Isomorphism
Kernel
Normal Subgroup
Quotient Group (Factor Group)

Techniques

1   To establish an isomorphism, find a homomorphism, and check that the kernel comprises the identity alone.

2   When specifying a homomorphism, make sure it is well-defined—if a group element can be written in more than one way, check that the specification of the homomorphism is independent of the way the group element is expressed.

3   When manipulating group elements, make use of expressions of the form

$$gxg^{-1}$$

because you can do things like:

$$g(xyz)g^{-1} = (gxg^{-1})(gyg^{-1})(gzg^{-1})$$

and

$$gh(x)h^{-1}g^{-1} = gh(x)(gh)^{-1}$$
$$gy = gyg^{-1}g$$
$$(gxg^{-1})^{-1} = gx^{-1}g^{-1}.$$

Postscript

"Map me no maps sir; my head is a map, a map of the whole world."

Fielding
Rape Upon Rape Act i, sc. 5

5.6   FURTHER SELF-ASSESSMENT QUESTIONS

SAQ 13

Let A be the cyclic group of order n generated by a, and B be the cyclic group of order 2 generated by b. That is,

$A = \langle a \rangle$ and $a^n = e$ (we identify a^{-1} with a^{n-1}, etc.).
$B = \langle b \rangle$ and $b^2 = e$.

Let D_n be the group generated by a, b with the relation

$$bab = a^{-1}.$$

This means that the elements of D_n can all be written in the form

$$a^i b^j, \quad \text{where } 0 \leqslant i < n, 0 \leqslant j < 2.$$

For the group operation, we have

$$a^i a^j = a^{i+j} \qquad \text{(we identify } a^{n+k} \text{ with } a^k \text{ etc.)}$$

$$(a^i b)(a^j b) = a^i a^{-j} = a^{i-j}$$

$$(a^i b)a^j = a^i b a^j b b = a^{i-j}b$$

$$a^i(a^j b) = a^{i+j}b.$$

D_n has order $2n$. For a physical representation, $n > 2$, consider a regular polygon with n sides. D_n is the group of symmetries: a represents rotation through $\dfrac{2\pi}{n}$ degrees, and b represents a reflection about one fixed axis of symmetry passing through a vertex and the centre of the polygon.

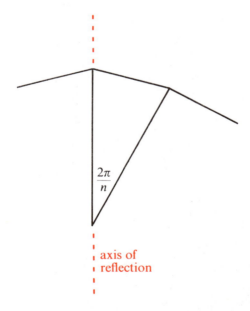

axis of reflection

(i)   Write out the group tables for $n = 3, 4$.

(ii)   Show that A and B are subgroups of D_n.

(iii) Show that A is a normal subgroup of D_n. Is B normal?

(iv)   Show that $D_n/A \approx B$.

(v)   Show that $T : a \longmapsto bab$ is a homomorphism from A to A. Is T an isomorphism?

(vi)   Show that $D_n = AB$.

(vii) What is the centre of D_n? (HINT: Look at the cases $n = 3, 4$ first.)

(viii) Show that D_2 is Abelian, and that $D_3 \approx S_3$.

(ix) Show that $\{e, a^i b\}$ is a subgroup of D_n isomorphic to B. How many subgroups of order 2 are there?

(x) Show that $N = \{e, a^3, a^6, a^9, a^{12}\}$ is a normal subgroup of D_{15}.

(xi) Describe D_{15}/N. (Find its order, show that it is isomorphic to a group you know.)

(xii) Find D_n'. (There are two cases: n even; n odd.)

(Solution is given on p. 42.)

SAQ 14

Let G be a group, N a normal subgroup of G, K a subgroup of G.

(i) Show that $N \cap K$ is normal in K.

(ii) Show that N is normal in NK.

(Solution is given on p. 45.)

5.7 SOLUTIONS TO SELF-ASSESSMENT QUESTIONS

Solution to SAQ 1

Let A be an Abelian group, and H be an arbitrary subgroup.

Since

$$aHa^{-1} = \{aha^{-1} : h \in H\} = \{aa^{-1}h : h \in H\} = H \quad \forall a \in A,$$

H is a normal subgroup of A.

Solution to SAQ 2

Problem 8

To show that, for fixed g, the set gHg^{-1} is a subgroup of G, we need only check that:

(i) $ghg^{-1}, gkg^{-1} \in gHg^{-1} \Rightarrow (ghg^{-1})(gkg^{-1}) \in gHg^{-1}$;

(ii) $ghg^{-1} \in gHg^{-1} \Rightarrow (ghg^{-1})^{-1} \in gHg^{-1}$.

(See ***Herstein***: *page 32, Lemma 2.3*.)

(i) $\forall h, \ k \in H$ we have

$$(ghg^{-1})(gkg^{-1}) = ghkg^{-1}.$$

Since H is a subgroup of G, $hk \in H$, and so $ghkg^{-1} \in gHg^{-1}$, i.e., gHg^{-1} is closed under group multiplication.

(ii) Also, $\forall h \in H, h^{-1} \in H$ and

$$(gh^{-1}g^{-1})(ghg^{-1}) = gh^{-1}ehg^{-1}$$
$$= geg^{-1}$$
$$= e.$$

Thus

$$(ghg^{-1})^{-1} = gh^{-1}g^{-1}.$$

Since H is a subgroup of G, $h^{-1} \in H$, and so $gh^{-1}g^{-1} \in gHg^{-1}$.

Hence gHg^{-1} is a subgroup of G.

Problem 9

To show that H is normal in G, observe that, for any $g \in G$,

gHg^{-1} is a subgroup of G.

Now all we know about H is that it is the only subgroup of G which has order $o(H) = p$, say. But gHg^{-1} is a subgroup of G, and it is easy to see that it has the same order as H, for

$$h \longmapsto ghg^{-1} \text{ specifies a function } T_{g^{-1}} : H \longrightarrow gHg^{-1}.$$

But $T_{g^{-1}}$ is one-to-one since $T_{g^{-1}}(h) = T_{g^{-1}}(k)$ implies that

$$ghg^{-1} = gkg^{-1} \text{ or } h = k.$$

Thus gHg^{-1} is a subgroup of G with order p. Therefore, by hypothesis, $gHg^{-1} = H$, and this must be true for all $g \in G$. Consequently, H is a normal subgroup of G, by ***Herstein***: *page 42, Lemma 2.9.*

Solution to SAQ 3

We shall show that, for all $g \in G$,

$$gHg^{-1} \subset H.$$

We are given that the product of any two right cosets is a right coset, so, in particular, if $g \in G$, then

$$HgHg^{-1} = Ht$$

for some $t \in G$. Which t? Well, we know that $e \in H$, so

$$egeg^{-1} \in Ht.$$

Thus, $e \in Ht$, so $Ht = H$, the only coset containing e.

Therefore, for any $g \in G$,

$$HgHg^{-1} = H.$$

In particular, this means that

$$egHg^{-1} \subset H$$

or that

$$gHg^{-1} \subset H$$

for any $g \in G$. This completes the proof.

Solution to SAQ 4

Let $g \in G - H$. Then the right cosets of H in G must be H and Hg. The left cosets must be H and gH. Now

$$H \cap Hg = \varnothing \qquad H \cup Hg = G.$$
$$H \cap gH = \varnothing \qquad H \cup gH = G.$$

We conclude that $gH = Hg$, whence by *Herstein: page 42, Lemma 2.10*, H is a normal subgroup of G.

Alternatively, H and $G - H$ must be the two left cosets, and the two right cosets.

Thus every left coset is a right coset and so, by *Lemma 2.10*, H is normal.

Solution to SAQ 5

(i) To show that NH is a subgroup, we need only show that $NH = HN$, and the result then follows by *Herstein: page 38, Lemma 2.8*.

The fact that N is normal means that

$$\forall h \in H, \quad hNh^{-1} = N;$$

thus

$$hN = Nh.$$

This means that

$$HN = NH$$

and the proof is complete.

(ii) First, by the definition of a normal subgroup, it follows that if N is normal in G, then N is normal in H. Thus, it is meaningful to talk about H/N.

To show that H/N is a subgroup of G/N, we show that it is closed under the multiplication in G/N and that it contains the inverses of all its elements.

(See *Herstein: page 32, Lemma 2.3*.) Both proofs are trivial.

Closure

For $Nh_1, Nh_2 \in H/N$,

$$Nh_1 Nh_2 = Nh_1h_2 = Nh, \text{ for some } h \in H.$$

Inverses

Let $Nh \in H/N$. Then $h \in H$, $h^{-1} \in H$.

Consequently,

$$NhNh^{-1} = Nhh^{-1} = N,$$

the identity in H/N. So, the inverse of Nh in N/H is Nh^{-1}. But $Nh^{-1} \in H/N$ as required.

The proof is now complete.

Solution to SAQ 6

By **Herstein**: *page 47, Lemma 2.13*, we know that the mapping ϕ from G to G/N defined by

$$\phi(x) = Nx \text{ for all } x \in G$$

is a homomorphism onto G/N. However, the homomorphic image of an Abelian group is Abelian because if $a, b \in \phi(G)$, then $a = \phi(x)$ and $b = \phi(y)$ so

$$ab = \phi(x)\phi(y) = \phi(xy) = \phi(yx) = \phi(y)\phi(x) = ba.$$

Solution to SAQ 7

(i) We must show that the subset $\phi^{-1}(\overline{H})$ of G is closed under group multiplication and the formation of inverses.

Thus, if x and $y \in \phi^{-1}(\overline{H})$, then

$$\phi(x) \in \overline{H}, \qquad \phi(y) \in \overline{H}.$$

Therefore $\phi(x)\phi(y) \in \overline{H}$ (because $\overline{H}$ is a subgroup of $\overline{G}$) which is the same as $\phi(xy) \in \overline{H}$.

Thus

$$xy \in \phi^{-1}(\overline{H}).$$

Also, if $x \in \phi^{-1}(\overline{H})$, then $\phi(x) \in \overline{H}$, so $\phi(x)^{-1} \in \overline{H}$ since $\overline{H}$ is a group.

But $\phi(x)^{-1} = \phi(x^{-1}) \in \overline{H}$ and so

$$x^{-1} \in \phi^{-1}(\overline{H}).$$

Thus $\phi^{-1}(\overline{H})$ is a subgroup of G.

(ii) The kernel K of ϕ is a subgroup of $\phi^{-1}(\overline{H})$ since $\overline{e} \in \overline{H}$ and so

$$K = \phi^{-1}(\overline{e}) \subset \phi^{-1}(\overline{H}).$$

(iii) As usual, we shall prove that $\phi(A)$ is closed and contains its inverses. First, if $a, b \in \phi(A)$, then $a = \phi(x)$, $b = \phi(y)$ for some $x, y \in A$. Thus,

$$ab = \phi(x)\phi(y) = \phi(xy) \in \phi(A),$$

since $xy \in A$.

Next, if $a \in \phi(A)$, then $a = \phi(x)$ for some $x \in A$. But since A is a subgroup of G, $x^{-1} \in A$ and so

$$a^{-1} = \phi(x)^{-1} = \phi(x^{-1}) \in \phi(A),$$

completing the proof.

We can summarize this result by saying that *the homomorphic image of a group is also a group.*

(iv) To show that $\phi^{-1}(\phi(A)) = KA$ we have merely to check that the two sets are equal. Notice that by (iii) and (i), the left-hand side is a subgroup of G. We know already that KA is a subgroup of G, since K is normal. Now, by *Lemma 2.16*, for any $a \in A$,

$$\phi^{-1}(\phi(a)) = Ka.$$

Thus,

$$\phi^{-1}(\phi(A)) = \{\phi^{-1}(\phi(a)) : a \in A\}$$
$$= \{Ka : a \in A\}$$
$$= KA.$$

Solution to SAQ 8

$\phi : x \longmapsto gxg^{-1}$, where g is a fixed member of G.

The fact that ϕ is a homomorphism is shown by:

$$\begin{aligned}
\phi(x)\phi(y) &= (gxg^{-1})(gyg^{-1}) \\
&= gx(g^{-1}g)yg^{-1} \\
&= gxyg^{-1} \\
&= \phi(xy).
\end{aligned}$$

$$\begin{array}{ccc}
(x, y) & \longrightarrow & xy \\
\phi \times \phi \downarrow & & \downarrow \phi \\
(gxg^{-1}, gyg^{-1}) & \longrightarrow & gxyg^{-1}
\end{array}$$

To show that ϕ is an isomorphism, we show that the kernel of ϕ is $\{e\}$. So suppose x is in the kernel of ϕ. Then

$$\phi(x) = gxg^{-1} = e$$

and so

$$x = g^{-1}(gxg^{-1})g = g^{-1}eg = e.$$

Thus, the kernel of ϕ is $\{e\}$ and ϕ is an isomorphism.

The fact that ϕ is onto G is easily shown.

For any $y \in G$,

$$\phi(g^{-1}yg) = g(g^{-1}yg)g^{-1} = y.$$

Thus for any $y \in G$,

$$\phi : g^{-1}yg \longmapsto y$$

and, therefore, ϕ is onto.

Solution to SAQ 9

As suggested, let

$$\phi : G \longrightarrow G$$
$$\phi : y \longmapsto y^n.$$

We shall prove as usual that

(i) ϕ is a homomorphism

(ii) ϕ is an isomorphism (the kernel of ϕ is $\{e\}$)

(iii) ϕ is onto.

(i)   Since G is Abelian, it follows from *Unit 4, Group Axioms* SAQ 10, page 18, that

$$x^n y^n = (xy)^n.$$

Consequently,

$$\phi(xy) = (xy)^n = x^n y^n = \phi(x)\phi(y).$$

(ii)   Suppose x is in the kernel of ϕ. Then

$$\phi(x) = x^n = e.$$

Also, we know that

$$x^{o(G)} = e.$$

But n is relatively prime to $o(G)$. Therefore, we can find integers λ and μ such that

$$\lambda o(G) + \mu n = 1,$$

so

$$x = x^{\lambda o(G) + \mu n} = e^\lambda e^\mu = e.$$

Thus $x = e$ and the kernel of ϕ is $\{e\}$.

(iii) Since ϕ is an isomorphism from G to G, ϕ is one-to-one from G to G. However, since G is finite, a one-to-one function from G to G is onto.

Thus for any $g \in G$, there exists $y \in G$ such that $g = y^n$.

Solution to SAQ 10

We used the fact that every subgroup of an Abelian group is normal. This proof will not carry through for non-Abelian groups for this reason.

Solution to SAQ 11

We have $\tau_{a,b} : V \longrightarrow V$ specified by

$$x \longmapsto ax + b,$$

and

$$G = \{\tau_{a,b} | a, b \text{ real}, a \neq 0\}, \qquad N = \{\tau_{1,b} \in G\}.$$

Before we show that N is normal in G, we should show that G is a group (under composition of functions which we know to be associative), and N is a subgroup. This is easily done since:

$$\tau_{a,b} \circ \tau_{c,d} = \tau_{ac,ad+b} \qquad \text{(closure)}$$

$$\tau_{a,b} \circ \tau_{1,0} = \tau_{1,0} \circ \tau_{a,b} = \tau_{a,b}$$

so $\tau_{1,0}$ is the identity, and

$$\tau_{1/a,-b/a} \circ \tau_{a,b} = \tau_{1,0} \qquad \text{(inverses)}.$$

Thus G is a group. Also, for N to be a subgroup we check:

$$\tau_{1,b} \circ \tau_{1,d} = \tau_{1,b+d} \qquad \text{(closure)}$$

and

$$\tau_{1,-b} \circ \tau_{1,b} = \tau_{1,0} \qquad \text{(inverses)}.$$

The normality of N can be checked in two ways. First, by direct calculation that

$$\tau = \tau_{a,b} \circ \tau_{1,c} \circ \tau_{1/a,-b/a} \in N.$$

The calculation is:

$$\tau = \tau_{a,ac+b} \circ \tau_{1/a,-b/a}$$

$$\tau = \tau_{1,-b+ac+b} \in N.$$

Alternatively, we can show that N is the kernel of some homomorphism, which is what we are asked to do in the next part. We must first specify a function from $G \longrightarrow (R - \{0\}, \cdot)$. This is the point at which insight is needed. The multiplicative part of $\tau_{a,b} \circ \tau_{c,d} = \tau_{ac,ad+b}$ is basically in the a, c, ac part. We therefore try

$$\phi(\tau_{a,b}) = a.$$

We can use the fact that we want N to be the kernel to help us in our choice, because G/N has the effect of neglecting the N-part of G, which is the set of elements of the form $\tau_{1,b}$. So we try concentrating on the remaining features, namely the a part.

ϕ is a homomorphism because

$$\phi(\tau_{a,b} \circ \tau_{c,d}) = \phi(\tau_{ac,ad+b})$$

$$= ac$$

$$= \phi(\tau_{a,b}) \cdot \phi(\tau_{c,d}).$$

The kernel of $\phi = \{\tau_{a,b} : \phi(\tau_{a,b}) = 1$, the identity of $(R - \{0\}, \cdot)\}$

$$= \{\tau_{a,b} : a = 1\}$$

$$= \{\tau_{1,b} \in G\} = N.$$

Therefore $G/N \approx (R - \{0\}, \cdot)$.

Solution to SAQ 12

To find the normal closure $\langle c \rangle^N$ in A_4, we use the group cards. Place c and c^2 as shown:

$$\boxed{c}$$

$$\boxed{c^2}$$

(c^2 is clearly in $\langle c \rangle$ and $\langle c \rangle^N$.)

Calculate gcg^{-1} and gc^2g^{-1} for the elements of g of order 3 first.

g	g^{-1}	gcg^{-1}	gc^2g^{-1}
ac	abc^2	abc	bc^2
abc^2	ac^2	bc	ac^2
bc	ac^2		
ac^2	bc		
abc	bc^2		
bc^2	abc		

Already we see that $\langle c \rangle^N$ contains $\{e, c, c^2, abc, bc, ac^2\}$.

Therefore it contains $abc \cdot c^2 = ab$

$$bc \cdot c^2 = b$$

$$ac^2 \cdot c = a.$$

And so $\langle c \rangle^N = A_4$.

An alternative approach: We know from past experience that A_4 has only one proper normal subgroup. Since that subgroup is $\{e, a, b, ab\}$ and does not contain c, the only normal subgroup of A_4 containing c is A_4 itself.

Solution to SAQ 13

(i)

D_3	e	a	a^2	ab	a^2b	b
e	e	a	a^2	ab	a^2b	b
a	a	a^2	e	a^2b	b	ab
a^2	a^2	e	a	b	ab	a^2b
ab	ab	b	a^2b	e	a^2	a
a^2b	a^2b	ab	b	a	e	a^2
b	b	a^2b	ab	a^2	a	e

D_4	e	a	a^2	a^3	b	ab	a^2b	a^3b
e	e	a	a^2	a^3	b	ab	a^2b	a^3b
a	a	a^2	a^3	e	ab	a^2b	a^3b	b
a^2	a^2	a^3	e	a	a^2b	a^3b	b	ab
a^3	a^3	e	a	a^2	a^3b	b	ab	a^2b
b	b	a^3b	a^2b	ab	e	a^3	a^2	a
ab	ab	b	a^3b	a^2b	a	e	a^3	a^2
a^2b	a^2b	ab	b	a^3b	a^2	a	e	a^3
a^3b	a^3b	a^2b	ab	b	a^3	a^2	a	e

Notice that $D_3 \approx S_3$.

(ii) A is a subgroup because it is just the cyclic group of order n, C_n. B is a subgroup because it is C_2.

(iii) To show that A is a normal subgroup of D_n, we *could* proceed to show that both

$$(a^i b) a^j (a^i b)^{-1} \in A$$

$$(a^i) a^j (a^i)^{-1} \in A$$

for each i and j, and after a somewhat lengthy calculation, achieve the desired result. As is often the case in group theory, however, the result is a trivial consequence of a more general result, namely SAQ 4, page 11.

Since

$$o(D_n) = 2n$$

and

$$o(A) = n,$$

A must have index 2 in D_n. Thus, by SAQ 4, A is normal in D_n.

The subgroup B, however, is not normal because, for example,

$$aba^{-1} = (ab)a^{-1} = (b^{-1}a^{-1})a^{-1} = ba^{n-2} \notin B.$$

(iv) Again, we appeal to a general result rather than plunging in and constructing an isomorphism.

Since

$$o(B) = 2,$$

and

$$o(D_n/A) = o(D_n)/o(A) = 2n/n = 2,$$

we shall achieve our desired result if we can show that *all* groups of order 2 are isomorphic. The latter fact is trivial: if (e, a) and $(\bar{e}, b)$ are two such groups, then the mapping

$$\phi : e \longmapsto \bar{e}$$

$$\phi : a \longmapsto b$$

is clearly a one-to-one homomorphism which is onto. Thus

$$D_n/A \approx B.$$

(v) Since $b^{-1} = b$, the mapping $T : a \longmapsto bab$ is of the form

$$T_g : a \longmapsto gag^{-1}$$

which we already know to be an isomorphism: $A \longrightarrow A$, since A is normal.

(vi) $o(A) = n$, $o(B) = 2$, $o(D_n) = 2n$. It follows that

$$o(AB) = o(A)o(B)/o(A \cap B) = o(A)o(B) = 2n$$

and so $AB = D_n$.

(We know from SAQ 5, page 11 that AB is a group because A is normal in D_n.)

(vii) For $n = 3$, the centre of D_n is $\{e\}$.

For $n = 4$, D_n has centre $\{e, a^2\}$: $\quad (a^2)(b) = a^2 b$

$$(b)(a^2) = a^2 b.$$

An easy calculation shows that

$$b(a^i b) = a^{n-i} = (a^i b)b$$

only when $n = 2i$.

Thus for: n even, $Z(D_n) = \{e, a^{n/2}\}, (n > 2)$;

$$n \text{ odd}, \ Z(D_n) = \{e\}.$$

(viii) The group table for D_2 is

	e	a	b	ab
e	e	a	b	ab
a	a	e	ab	b
b	b	ab	e	a
ab	ab	b	a	e

showing that D_2 is Abelian.

The isomorphism $D_3 \approx S_3$ is given by: $a \longmapsto \psi$

$$b \longmapsto \phi.$$

(ix) $(a^i b)(a^i b) = a^{i-i} = e$. Thus $a^i b$ has order 2, and since all groups of order 2 are isomorphic,

$$\{e, a^i b\} \approx B.$$

For each i, $0 \leqslant i \leqslant n - 1$, we have a distinct subgroup of order 2, thus there are at least n of them. If n is odd, that is all. If n is even, $\{e, a^{n/2}\}$ is the only other one.

(x) We find $(a^i b)^{-1} a^{3j}(a^i b) = (a^i b)^{15-i}(a^{i+3j}b)$

$$= a^{15-3j} \in A,$$

and, of course, $(a^i)^{-1} a^{3j}(a^i) = a^{3j} \in A$. Thus N is normal in D_{15}.

(xi) $o(D_{15}/N) = \dfrac{30}{5} = 6.$

A reasonable group might be D_3. For clarity, we shall write D_{15} in terms of x, y, not a, b.

Then specify

$$\phi : D_{15} \longrightarrow D_3$$

by
$$
\left.
\begin{aligned}
x^{3i+1}y &\longmapsto ab \\
x^{3i+2}y &\longmapsto a^2b \\
x^{3i+3}y &\longmapsto b \\
x^{3i+1} &\longmapsto a \\
x^{3i+2} &\longmapsto a^2 \\
x^{3i} &\longmapsto e
\end{aligned}
\right\}
\quad
\begin{aligned}
&x^i y^j \longmapsto a^\alpha b^j \text{ where } \alpha = \text{remainder upon dividing} \\
&\qquad\qquad i \text{ by } 3.
\end{aligned}
$$

ϕ can easily be shown to be a homomorphism.

Also $N = \text{kernel } \phi$. Thus $D_{15}/N \approx D_3$.

(xii) $aba^{-1}b^{-1} = aa = a^2$.

Thus $\langle a^2 \rangle \subset D'_n$.

Now D_n/A is Abelian, since it is of order 2, so $D'_n \subset A$.

There are two cases, n odd and n even.

If n is odd, $\langle a^2 \rangle = A$, since for some λ, μ, $a^{2\lambda+n\mu} = a$, which generates A.

If n is even, $\langle a^2 \rangle$ is a subgroup X of A of order $n/2$. We show that X is normal:

$$b(a^{2i})b^{-1} = a^{-2i} \in X,$$
$$a^j b(a^{2i})(a^j b)^{-1} = a^j ba^{2i}ba^{-j} = a^{-2i} \in X.$$

Thus X is normal.

We know that $X \subset D'_n$. We now look at D_n/X, and show that it is Abelian.

D_n/X has order $\dfrac{2n}{(n/2)} = 4$, and so must be Abelian.

Therefore by (ii) (c), page 30, $D'_n \subset X$. Consequently, $D'_n = X$.

Solution to SAQ 14

(i) Let $x \in N \cap K, k \in K$.

Then $k^{-1}xk \in N$, since N is normal in G and $k^{-1}xk \in K$, since $x, k \in K$.

Thus, $k^{-1}xk \in N \cap K$ and $N \cap K$ is normal in K.

(ii) NK is a subgroup by SAQ 5, page 11. And N is normal in G, so N is normal in any subgroup containing N.

TOPICS IN PURE MATHEMATICS

1	S	Set Axioms
2	S	Set Constructions
3	S	Sets and Numbers
4	A	Group Axioms
5	A	Group Morphisms
6		NO TEXT
7	T	Metric Space Axioms
8	T	Continuity and Equivalence
9	C	Finite State Machines
10		NO TEXT
11	A	Automorphism Groups
12	A	Group Structure
13	T	Topology Axioms
14	T	Topological Closure
15	T	Induced Topologies
16	C	Turing Machines
17	A	Rings and Ideals
18		NO TEXT
19	A	Special Rings
20	N	Categories
21	C	Recursive Functions
22	N	Universal Mappings
23	A	Euclidean Rings
24	A	Polynomials
25	C	Theory of Proofs
26		NO TEXT
27	T	Connectedness
28	T	Compactness
29	T	Fundamental Groups
30	T	Fixed Point Theorems
31	A	Field Extensions
32	A	Splitting Fields
33	A	Galois Theory
34	A	The Galois Correspondence

The letter after each unit number indicates the textbook required for that unit.

Key	S	Halmos	A	Herstein	N	not based on one of the texts.
	T	Mendelson	C	Minsky		

"Cards were at first for benefits designed,
Sent to amuse, not to enslave the mind."

David Garrick

Epilogue to Ed Moore's Gamester

GROUP CARDS

Introduction

At the back of this booklet, you will find a set of 52 group cards. The set comprises:

12 cards with 6 dots and 3 lines on each, labelled S_3 (6 red, 6 black);
24 cards with 8 dots and 4 lines on each, labelled A_4 (12 red, 12 black);
16 cards with 16 dots and 8 lines on each, labelled Q (8 red, 8 black).

These cards provide physical representations of three groups (S_3, A_4, Q). They will enable you to test the abstract ideas developed in the correspondence texts on concrete examples of groups which you can literally hold in your hands. You will find the cards useful when working on *Units 4, 5, 11* and *12*; they will enable you to perform calculations without resorting to a group multiplication table.

Labelling the Cards

Each card represents a group element and consists of a set of left-hand dots joined by paths to a set of right-hand dots. The group operation is defined by juxtaposition. The product of two group elements is determined by juxtaposing the two cards and examining the composite paths so formed. The product is represented by the card whose paths are the same as the composite paths.

Example 1: The Group S_3

From the six black cards marked with S_3, choose the following:

 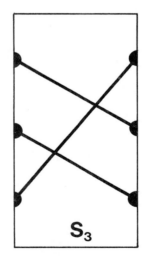

Label the first one ϕ and the second one ψ, at the centre of the top of the card as shown below. Now juxtapose the two in the order $\phi\psi$:

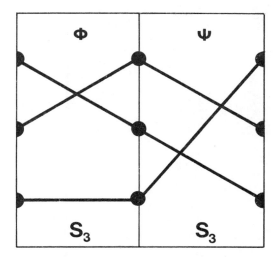

Choose the black card which summarizes the juxtaposition. The top left dot is joined to the bottom right dot, and the middle left dot is joined to middle right dot, so the "product" is:

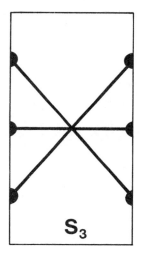

Label this card $\phi\psi$.

Repeat the procedure in the order $\psi\phi$, labelling the new black card.

Choose the card which has the effect of the identity, and label it e.

One card remains unlabelled. Juxtapose it with ψ, in both orders, and observe that the result is e in each case. Label this card ψ^{-1}.

You should now label the red S_3 deck in the same way as the black deck. The red deck provides a second copy so that calculations such as squaring an element can be carried out. Multiply the red ψ by the black ψ and observe that the result is ψ^{-1}. Notice also that ψ and ψ^{-1} are mirror reflections of each other, as they must be, since each is the inverse of the other.

Example 2: The Group A_4

Label (in both the red and black decks) the identity as e, and the three cards below as shown:

 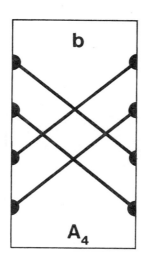 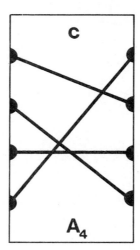

Now calculate and label each of the products

$$ab, \quad ac, \quad bc, \quad abc,$$

$$c^2, \quad ac^2, \quad bc^2, \quad abc^2.$$

Note that c^2 can be computed by using a red and a black copy of c.

Now that all the elements of A_4 are labelled, you can use these labels to short-cut calculations. For instance:

$$(ab)c = a(bc) = abc$$

and

$$ac^2 = (ac)c.$$

Example 3: The Quaterions

The cards have already been labelled, but you should check that the labels

$$a^2, \quad a^3, \quad a^3b, \quad b^3$$

are consistent with the cards labelled a and b. Observe also that

$$a^2 = b^2 \quad \text{but} \quad a \neq b \quad \text{and} \quad a^3 \neq b^3.$$

Using the Cards

You should now satisfy yourself that each of the packs S_3, A_4 and Q represent groups. To do this, first practise computing products by finding the inverse of each card. This can be done by looking for symmetry (and in some cases a card is its own inverse). Next satisfy yourself that the juxtaposition of any two cards from one pack is represented by some card of the pack. Since the packs are closed under formation of products and inverses, they form a group.

To gain facility with the cards, practise solving equations of the form

$$xy = z,$$

where two of the cards x, y, z are chosen randomly from one pack, and your job is to find the third. Make use of inverses when solving such equations. Further uses of the cards will be indicated in *Units 11* and *12*.